J. W. Bradley, T. G. Goodwin

A Manual of Illumination on Paper and Vellum

J. W. Bradley, T. G. Goodwin

A Manual of Illumination on Paper and Vellum

ISBN/EAN: 9783744642194

Printed in Europe, USA, Canada, Australia, Japan

Cover: Foto ©berggeist007 / pixelio.de

More available books at **www.hansebooks.com**

A Manual of Illumination.

On Paper and Vellum.

By J. W. BRADLEY, B.A. and T. G. GOODWIN, B.A.

NINTH EDITION,

CAREFULLY REVISED AND MUCH ENLARGED, WITH PRACTICAL NOTES,
AND ENTIRELY NEW ILLUSTRATIONS ON WOOD.

By J. J. LAING.

Ars probat artificem.

LONDON:

WINSOR & NEWTON, 38 RATHBONE PLACE,

*Manufacturing Artifts' Colourmen, and Drawing-Paper Stationers, by Appointment to
Her Majefty, and to H.R.H. the Prince Confort.*

Contents.

Part I.

ILLUMINATION 1

Part II.

MATERIALS 24
    Colours. 25
    Inks 28
    Vellum, Boards, and Papers. 29
    Pens 30
    Brushes. 31
    Burnisher and Tracer 31
    Metallic Preparations, &c. 32

Part III.

HOW TO SET TO WORK 33
    Ruling for Margins and Lettering 36
    Border. 37
    Initials 38
    Text or Lettering 39

Part IV.

COLOURING 41
    Table of Colours and Mixed Tints 43
    Diapers 50
    Line Finishings 52

Part V.

GILDING 53
ADDITIONAL MEMORANDA 60

Appendix.

Landscape 64
Figure 65
Styles 68
Copying 72
Design 74

Practical Notes.

Paper or Vellum 82
Sketching and Tracing 82
Transferring 83
Colouring in Masses or Grounds 84
Burnishing 87
Detail Colouring 88
Design 95

𝔓reface

THE NINTH EDITION.

INCE the publication of the first edition of this work, great advancement has been made in the practice of the Art of Illumination:—an Art daily becoming better understood and more appreciated.

Careful revisions of this Manual enabled succeeding editions to keep pace with this progress. In this, however, the *ninth* edition, still more important additions and alterations have been made. Not only has it been fully revised, but "Practical Notes," and other useful matter has been added; and a "Companion," full of choice illustrations, materially assists in explaining the art thoroughly, not only to the beginner, but to the more advanced student.

Woodcuts have been adopted in preference to lithographs, or chromo-lithographs, as rendering more truthfully the firmness of drawing and finished execution that distinguish true Illumination. These have been carefully selected and drawn on the wood by the reviser and faithfully engraved by the Misses Byfield.

A

MANUAL OF Illumination

on

Paper & Vellum.

𝔓art 𝔉.

ILLUMINATION.

N a Manual which is intended to be a concise practical guide on Illumination, a brief sketch of the art as it has flourished in bygone ages, will be found not only interesting, but to a certain extent most requisite, so as to enable the reader to better understand the points and purposes of our practical instructions. In the outset at the same time, it is essential that students should examine carefully the styles of illumination from such original manuscripts, as are exposed to the public view in our museums; and determine on painting exactly and truthfully choice parts from them, so as to gain by patient copying a practical acquaintance with their methods and details. For it is only by such painstaking that knowledge and perfection in this, as in all arts, can be attained, and difficulties overcome—moreover it is indispensable to proficiency in design. By this means, the student may also trace the excellencies and changes of one stage to another, each forming a link in the development of the art until it reached its greatest glory at the end of the xiv century. From the xv century, the art changed to a very different style, and gradually became somewhat debased until printing came into use.

Each style has its admirers, some delighting in the remote Anglo-Celtic, some in the Byzantine, others again in the Gothic, or about from the xiii to xiv centuries, (acknowledged by the ablest judges to be the best period,) and most of all probably in the Renaissance, dating from the xv century downwards to the present day of its attempted revival.*

---

* In regard to this, it may be mentioned as most important that this should be a *true* revival. Much injury has been done by charlatanism having undertaken to teach the art, and by the production of examples after no ancient manner, unless it be from the worst periods. In *reviving* an art, we must go back to the point at which it began to fail; and must humble ourselves to copy first, to become acquainted with its elements, before we can design well, as the artists in architecture, glass-painting and illumination of one period did from those of the former, as any one may trace; and by this obedience improved step by step, until the xiv century. But in the xv century, they stopped this succession of obedience, and reversed the former ways in colours and everything possible for attempts at originality—then, art decayed and fell. There are great laws of successive obedience in nature tending to gradual developments of perfection, that cannot be infringed without hurt; and why not in art that has been a necessary part of man's existence in all ages, and held on progressing most truly in every branch to the xiv century? M. Paul Durand, one of the most learned antiquaries and artists of France, who has travelled repeatedly in European and Eastern countries, for close study of art, traces a gradual connecting progress from the Egyptian artists down to the end of the xiii century. If you look very closely at the beautiful work on metal bowls and frescoes found at Nineveh by Mr. Layard, placed in the British Museum, you will see some resemblance to the finishings of the xii and xiii illuminated MSS. M. Digby Wyatt, in writing of the originality of the Anglo-Irish MSS. as being very decided, yet says, "in pose and motive it is generally obvious that some ancient model has been held in view." Since this rupture of disobedience (now some four hundred years), we have been floundering about, making futile attempts at style after style, classicalisms

One great use of this small Manual will be, to assist in explaining in the simplest manner the characteristics of a given style, thus enabling students on seeing a piece of illuminated work, to determine with tolerable precision the period to which it belongs; also thereby in their own practice to avoid many errors; but to do this without confusing, many peculiar styles such as the Oriental and Moorish must be left unconsidered.

In Europe, then, we look for the earliest available examples to the later Roman or Greek Empire, and the reign of the Emperor Justinian. Their peculiar features are—rose or purple-stained vellum, and the lavish use of gold and silver. The style is based on the traditions of classic periods of art, borrowing the heavy enrichments and semi-circular arches of the best Roman work, and adding thereto sundry mystic symbols of a later growth. The oldest existing examples of Roman art are attributed to the III or IV century—perhaps the oldest that contains pure ornament is a Roman Calendar at Vienna. The Gothic ' Gospels' preserved at Upsal, and called by way of eminence ' codex argenteus'—*i. e.* written on purple in [gold and] silver letters, is an example of this ancient class. It dates about A.D. 360.

A copy of Homer, in gold and purple was presented by a Greek Empress to her son, early in the III century; and doubtless by no means an uncommon gift, for we

after classicalisms, and trying new orders of architecture, all ending most unsatisfactorily, and at last are driven back to the XIII and XIV century models for architecture and glass-painting. May not this hold good as a reason for thinking we may have to go back to those periods for the sister art of illumination? May not these laws of successive obedience be the great " Free-Mason secrets " we have been lamenting the loss of? It is but natural that by disobedience we should lose the secrets of gold for illuminations, and glowing colours such as the rubies for glass-painting. Such statements as these may appear unnecessary here for a beginner, but it is important in the present state of the arts, that students should have rules for guidance to enable them to start right, and we wish to impress upon them that they must *copy* the XIII and XIV century examples, before they can hope to produce a good XIX century style.

read that by the time of Leo the Great, the art was popular enough to support a separate and superior class of artists. One reason that these MSS. are now so rare, occurs in the fact, that in the year 470, the said Leo restricted by sumptuary edict, the use of purple to himself and his imperial successors.

After the reign of Commodus, art began rapidly and grievously to decline, especially in the West. Intercourse with Asiatic artists, however, had kept it alive in the Greek Empire, and from time to time skilful Greeks had been invited to the court of Persia. Hence, when Justinian set on foot the great works at Constantinople, these travelled Greeks were recalled, and artists from various parts of the East invited to join them in the undertaking. This gave to indigenous talent a new life. Gold, silver, and mosaic—Eastern inventions—combined with Western types to produce a style that is hence for-ward called Byzantine.

When the Imperial court was transferred from Rome, art also went with it to the new capital, and as far as books were concerned, flourished there. Now we find it stretching back again over Europe, and meeting with patronage and encouragement again at Rome. In the VII century, a costly treasure in the shape of books —rejoicing in the rich splendour of Byzantine art, and a marvel to all beholders—was bestowed upon the cathedral at York. It was placed there by Bishop Wilfred on return-ing from his travels. With Wilfred also came over certain foreign artists, whom he had selected for their skill in glass-work and illumination, and whose arrival, no doubt, gave a fresh impulse to the enthusiasm of native craftsmen.

In very ancient times, the Greeks are said to have learned many things of the Celts, who possessed curious arts, together with the most extensive knowledge of natural science. Those very remarkable knotted and intricate traceries of twigs and stems, met with so often on Celtic monuments and in certain old MSS., are of their invention and among their significant secrets.

The Celtic style—as we may conveniently term it—distinguished by its interlacing bands and inextricable coils of lizards, birds and twigs, was carried from Ireland by St. Columba to Iona, and thence to Lindisfarne by Aedan, when made bishop of that see in 735. From this time it is called Hiberno-Saxon, or Anglo-Hibernian. One noted example of Celtic work exists in the great 'Book of Kells' at Dublin, in which the peculiar niceties of the art are so elaborate as to make them the wonder and despair of copyists. "Of this book," says Mr. Wyatt, in his recent valuable Essay, "Mr. Westwood examined the pages, as I did, for hours together, without even detecting a false line, or an irregular interlacement."*

Of Hiberno-Saxon work, the example of greatest note is the famous "Durham Book" in the British Museum (Cott. Nero. D. IV.) This venerable MS. was produced in the course of the VIII century at Lindisfarne; being the votive offering of two successive bishops of the diocese, to the memory of St. Cuthbert. It is sometimes called St. Cuthbert's Gospels.†

* It would scarcely be possible or advisable for the most skilled draughtsman in modern days to copy some parts of such a book. The manner of life in the middle ages (ignored as *dark ages*) mainly contributed to the production of such unrivalled marvellous works of art. The way and rate at which we live in the XIX century is much against the practice and revival of those almost lost arts, unless the student works diligently and quietly in his 'scriptorium,' as they did of old. Many people very ignorantly condemn the revival of illumination, as needless, now that printing exists, but they may rest assured that its practice is—especially for young people—"most disciplinary and delightful, and tends, even as an accomplishment, to strengthen those qualities of *patience, thoughtfulness,* and *delicacy* which shed so salutary an influence upon our daily life." The school-boy or girl might very profitably and pleasantly acquire their first knowledge in free-hand drawing and colouring by heading their copy-book pages with an occasional simple capital at first, to relieve the tedium of the repeated text or small hand.

† As it is a very precious book, and some examples from it were published some years ago, which may be had at the print shops, it should be seldom consulted unless for important reference.

The Celtic style of ornament, after its introduction
into this country, became very popular, and influenced,
by degrees, the continental schools. We find the intricate tracery combined with the half classic Byzantine in
the examples handed down to us from the time of Charlemagne, both of French and Italian execution.

By the IX century, that is the reign of Charles the
Bald of France, the taste for gold letters and ornaments
on purple grounds, became very general. But books of
the same or a similar style, had reached England still
earlier. In 596, Augustine the missionary had come
over to romanize the Saxons, and had been supplied
with various books from Italy. Soon after his arrival
a *scriptorium* seems to have been founded, perhaps in
Kent: and Saxon artists taught what may be called an
Anglo-Roman style—Roman draperies being imitated in
the miniatures, and the uncial or rustic letters adopted
in the text.

In the time of Alfred, there was a *scriptorium* at
Winchester, founded, it is said, by St. Swithin, who
was made Bishop of Winchester in 852, about two
years before Alfred—a boy of five years old—went with
his father to Rome. The saint died in 863, and in 871
Alfred became king. On his return from Rome, Alfred
had seen at Paris the magnificent library of Charlemagne;
and when he became king, he did not forget the attrac-
tions he then met with. He founded another monastery
near the old one at Winchester, attaching of course a
scriptorium. This was the afterwards celebrated New
Minster. The singularly spirited and beautiful class of
work produced in the two schools at Winchester might
almost be termed an invention, indicating as it did so
vast an improvement or at least advance, upon the continental work of the preceding century. Yet it was really
like the rest, an importation; and grew naturally out of
the style practised the century before, at Paris and
Limoges, as may be seen by comparing it with the

ornamentation of the Bibles of Charles the Bold and Lothaire, in the Imperial Library at Paris. It was in use probably, still earlier in the Monastery of St. Augustine at Canterbury.

The book oftenest quoted as a specimen of this 'Opus Anglicum,' as Winchester work has been called, is the Benedictional belonging to the Duke of Devonshire. It was written by Godemann, a monk of the *Old* Minster or Priory of St. Swithin's, Winchester, for the Bishop St. Æthelwold, who was consecrated in 963, and died 987. Another production, it is thought, of the same *scriptorium* is the magnificent purple 'Gospels,' now at Stockholm. The Benedictional of St. Æthelwold is written in a clear Roman hand—capitals being in gold; alternate lines in gold, red, and black, sometimes occurring on the same page; the gold is executed with leaf, laid upon size, and afterwards burnished so as to appear solid and brilliant. It contains thirty illuminations, which with thirteen other pages, are surrounded with profusely ornamented borders, composed of bars, arches, &c., as in the Caroline MSS. but intertwined rather than inlaid, with foliage in various colours, firmly and richly painted.

Of the rival *scriptorium* at New Minster, perhaps the finest example is the 'Golden Book' of Edgar dating 966.* The Benedictional of Archbishop Robert, at Rouen, is another. This MS. presents a style so strikingly similar to that of the 'Æthelwold' as to induce some good judges to attribute it to the same hand. It is, if anything, bolder and more beautiful. The miniatures of this style are remarkable for their superior drawing to other English MSS. of the time, as for example the Bodleian Caedmon; and for the fluttering character of the dra-

---

* Called 'golden,' because the text is in raised gold. Title-page contains figures with flowing drapery, painted on a pink-salmony ground; other pages have a dry cobaltish-blue ground.

peries—so evidently the result of studies in classic or
nearly classic art. In some we are struck with their
similarity to the paintings preserved at Pompeii—there is
the rippling and zigzag edge and long Etruscan folds
observable, for instance in the winged Victory of the
House of Dioscuri; which though Roman, is imitated
from earlier examples as much Etruscan it may be as
Greek. The same management of drapery may be seen
in early Greek sculpture. In the schools of the VIII
and IX centuries, established by Charlemagne, "it is
presumed," says Sir F. Madden, "that Italian or German
artists (who worked after the models of the Greek school)
were chiefly employed."

We have now, proceeding from the Byzantine, shown
how it met and combined with the Celtic, to produce
the style prevalent at the close of the X century.
Both now became subordinate to the inventions that were
to exceed in boldness and skill all that had gone before;
not, we say, in splendour or elaboration, but in exquisite
beauty. One great characteristic of MSS. dating from
the VII century, particularly in the XI, and early XII
centuries, was the fancy for monster initials; the first
letter or word of a chapter sometimes occupying the
whole length, and even surface of the page, as in the
'Durham Book' and later works.* In Italy this fancy
prevailed to the very last, as may be seen in the choral

---

* The frontispiece illustration, Plate I, arranged from a title-page of
a large folio Bible in British Museum is an example of style of border
of XII century, reduced considerably less than one half real size. In the
medallions are subjects from the creation and fall. Our Lord is en-
throned in the centre within an aureole or "vesica piscis," the sacred
glory round the Divinity resembling a fish, which is drawn by the inter-
section of two circles. The Latin I and N are interwoven by foliage
and figures with the border, too intricate to give truly and usefully on a
small scale, but adapted for simple practice in the sequel sheets with
some omissions, as all the illustrations are; but the Christ is drawn
nearly original size to shew the manner of figure drawing and the flowing zigzag edged drapery. Christ is usually surrounded by the four
evangelists, whose symbols are here inserted—Saint Matthew being

books of the cathedrals, and other MSS. But in the xii century, the initials for the most part gradually diminished, giving way to the borders, which grew in richness; until in the xv, they became altogether detached from the constructions of the page, and were set in panels of sweet colour or bright gold among the text. Illumination then passed into miniature painting, the limner's art—limner being a term derived from the older and longer name of illuminator.

There is an interesting fact with respect to styles, that may be noticed here, it is their relation to the architecture of their respective periods. In Byzantine MSS. the tympana of the arches were often inlaid with tesselations of brilliant colour, like the mosaic on the walls of the Duomo of St. Mark, at Venice. In English and other MSS. of the xiii century, the backgrounds of the initials and miniatures were diapered or inlaid with colour and burnished gold.* In all these cases, the features are also

always represented by an angel or winged man with a book, standing or kneeling; Saint Mark by a winged lion; Saint Luke by a winged bull; and Saint John by an eagle, each having a plain nimbus or glory round head, Christ alone having the cross on nimbus. To employ these and other emblems properly, the student should read Didron's "Iconography of Christian Art," (Bohn, Covent Garden) in which he will get many illustrations suggestive for illumination, as Figs 58, 40, 67, 69, 74, &c., and Mrs. Jamieson's "Sacred and Legendary Art," (Murray). The initial letter I, page 1, drawn one-third real size, also from a folio Bible of xii century in British Museum (Harl. 2798-99) affords a sufficient example of the foliaged and figured great lettering; it commences the 'IN PRINCIPIO' of St. John's Gospel, the IN being the two principal letters, but not intertwined; the *principio* is arranged down side of the N, as in Plate I, but within, not on the border—the whole, with figures of Christ and St. John at top, forming a large painting nearly whole size of page. This MS. is generally open at this page at the Museum. Letter S, heading Part II, is xii century date.

* In Plate II, is shown treatment of architecture in xiii century, and therein is also gained some idea of the furniture of this period. Other examples such as I in Plate IV, with Ruth and infant below, and Boaz above, have canopied gables and turrets. These occur, also, most exquisitely designed in xiv century work, as on page 71 from Addl. MS. 17,341 British Museum.

architectural.* Diapering, however, was the fashion
of the day, extending to dress, furniture, horse-
trappings, and in fact everything to which it could
possibly be applied, and always with great variety and
effect.

From the XII to the XIV century is called the
' outline' period, owing to the immense improvement
in the spirit and contour of the figure drawing, which
being no longer confined to mere ornamental lines, as-
sumed expression as well as accuracy. By the middle of the
XII century, the heavy curling foliage of which the
Germans are still so fond, was exchanged generally for
lighter sprays of hawthorn, saint-foin, and ivy, convention-
ally treated—that is subordinated to the necessities of
construction, which is the true purpose of ornament; and
the interstices filled with pale yet brilliant tints of
colours.

It is this period that is notable for the introduction of
curious *figuræ monstrosæ*, *vulgò* grotesques, into various
parts of the decoration, at first painfully distorted but
afterwards possessing singular grace, sometimes sweeping
their lacertine tails for inches down a border, now and
then coiling up into an initial, and again quaintly budding
out at the extremity into ivy leaf forms. Besides these

---

* From X to the XIV centuries, illuminated MSS. display the same
artistic feeling or broad harmony we see in architecture with its sculpture and painted glass. This is found especially in the XII century work
of the French Cathedrals, such as Chartres, Poitiers, &c., where the
same solemn majesty of countenance and severe but graceful drapery is
expressed as in such MSS. as Cott. Nero. C. 4, British Museum, in
which the 'Jesse,' Raising of Lazarus, and Christ's Entry into Jeru-
salem, are good studies, and which with the rich canopies are most
suggestive to the student, (glass painters especially,) for XII century
designs. It may as well here be stated that to become a right good
illuminator, students should look well at Gothic architecture and glass
painting, and, if possible, (ladies as well) get the *entrée* into an architect's office, where good architecture, cartoon drawings, and even embroidered stuffs and furniture are designed ; there you may pick up
much in a few weeks, useful in other ways besides. You should pay for
the privilege, rather than for indifferent teachings.

are found delicate pen-drawings of various animals, such as dogs and hares, very quaint but exquisitely natural. This style deserves great attention, advancing as it did to perfection until the end of the XIV century, when schools became greatly multiplied in France, Italy and the Netherlands and prevailed at the same time in Persia and Hindostan. It should be thoroughly studied for its mastery over *expression*, and its firm yet extreme delicacy of handling points are to the illuminating artist of the greatest importance.

Amongst the best examples in the British Museum of the XII century which might be recommended for study, are Cotton, Nero, C 4; Harl. 2798-99;* Add. 14,788-† 89-90;‡ Royal, 10, A 13;§ perhaps verging on to the XIII century. Of the XIII century MSS., I. D. X. contains some good simple examples for the beginner in diapers,‖ small capitals, (generally gold set in blue or light red purple, with scarlet edging,) and in finishings at ends of sentences in blue and vermilion. Addl. 17,868, affords pure models of XIII century architecture for illumination, glass-painting, and to some extent for architects also. It is rich in delightful grotesques and line finishings, such as at end of Part II. Addl. 15,452, a most exquisitely finished MS. has, like most all good MSS., a fine "*Jesse*" design, with David, the Virgin, and Our Lord, supported by prophets in the encircling medallion branches. The head of Christ is surrounded by (amidst green leaves) the seven spirits, who denote,

* Initial letter 'I,' page 1, is from above MS.
† Frontispiece is arranged from this MS. principally.
‡ Letter 'S,' Part II. is from this volume.
§ Fig. 1, Plate II., is a good example from it of St. Dunstan. The saint is robed in blue, embroidered with gold; the inner ground or panel is brightburnished gold, then red stars and delicate white trefoils on pale purple border, then narrow green band, then white varied ornament on blue border, then gold, then green outer band.
‖ No. VII of diapers is from it.

Wisdom.	*Sapientia.*
Intelligence.	*Intellectus.*
Counsel.	*Consilium.*
Strength.	*Fortitudo.*
Knowledge.	*Scientia.*
Piety.	*Pietas.*
Fear.	*Timor.*

the last meaning "the fear of the Lord." The text of this MS. is written in a very small, firm, delicate hand. 11,639 is a Hebrew MS. containing some fine little pieces of colour in gold, blue, fine purples and scarlets, used effectively for borders.* The drawing throughout, unites boldness and exquisiteness of finish (especially in animals such as the lion and the horse). In many points this MS. may be an authority for artists for subjects connected with the ancient ceremonial law.

The Vegetius, 11,698, British Museum, is of all work, towards end of the xiii century, the most famous for the highest powers of firm graceful design, lively drawing, and pure rich colouring. There are no miniatures in it, all are splendid bold capitals—so very dignified and free, that many style them as *stiff.* As to colouring, the gold dazzles the eyes, and the purples are most glowing, sometimes cloudy with carmine, sometimes of a dry salmon vermillion tone.† Plate III.‡ and ' U,' Part III.,

---

* Mid border in Plate IX., and tail piece to end of Part III. is from this Hebrew MS.

† In speaking of *purples*, readers may be a little perplexed at first, when they would be inclined rather to call them *reds ;* but the ancient purples were not like ours now-a-days, which possess a preponderance of blue. The purples of illuminated MSS. vary so much, and are so mysterious to compound, that sometimes it is necessary to note them as being of a vermillion-purple tone, other times as of a pink-salmon flesh look, or of a carmine-vermillion tone, or clouded with glossy carmine, or lake, or heightened with a colour *like* Saturnian red, or gamboge and orange.

‡ The ' P ' shortened in Manual is complete in sequel illustration : ' T ' and ' Q,' subordinate capitals, occur in the grandly coloured title-page, the first lines of which are written in gold, (like Vegeti at bottom of plate,) in alternate lines on blue and cloudy purple with

are from this MS. as also some others in sequel illustrations. I. D. I., and 16,410, are good specimens of late xiii century work, somewhat similar to a very complete and highly prized MS. Bible of very late xiii century, in the Advocates' Library at Edinburgh, from which latter, Plates IV. and V. are drawn, together with the fine ' B,' heading Part IV. and other capitals in sequels,* described hereafter.

In this last-mentioned MS., one cannot but admire the expression of the faces and grace of the figure drawing. The faces of the prophets and apostles look wise and solemn, with lofty foreheads and wavy locks, and kings look like kings, although painted often in the space of less than an inch, and the heads no more than an eighth

gold bars between. Letter 'I' is a favourite mode of making a capital serve as a border (also complete in sequel). Leafage is all carefully drawn and varied, as shown on drawing, as in Fig. 2, (Plate III.) enlarged four times real size. Griffin's heads are all well expressed, sometimes as wisely-cunning, sometimes as very gentle, and sometimes as if they would split their jaw-bones. Finishings of sentences are drawn as in Fig. 1, and in title page like tail pieces, end of Part IV. Note the eager expression of the hunter with his hair on end, and the drawing of the animals. Some parts of the designs of "Idylls of the King," by F. S. A. are based on Vegetius, and other choice examples.

* In mentioning the sequel illustrations so often, it is as well to explain at once that they form part and parcel of the plan of this present new edition of Manual, and are all arranged, and in some parts altered so as to form a series of the best practical examples in xii, xiii, xiv, and even xv century work for beginners and students. A set or two of them as printed from the blocks, will be hand-coloured from the originals, for the student to refer to in colouring theirs, without going to the British Museum, where space is limited for illuminators, and for access to the valuable illuminated MSS. which mere learners should not expect at first to study, until they obtain some mastery over pencil, pen, and brush handling. Copy first from some good missal in possession of a friend, or from some of the best authentic published examples by Mr. M. Digby Wyatt, Mr. Shaw, Mr. Noel Humphreys, and Mr. Timbs ; only it must be observed that chromo or coloured lithographics cannot give the real colours nor careful drawing of the originals, so you must always try and do beyond them in proper colour and finish. When you peruse an illuminated MS., be particular not to turn the leaves over in masses, but singly, and have no pencil, pen, or brush in your hand at the time.

of an inch. In the 'B' already mentioned, commencing the "Beatus" of the Psalms, the whole *pose* of David, and expression of head which is but a tenth of an inch, gives the idea of the exultant boy asking Goliath 'where are thy Gods now?' and Goliath seems mightily humbled. More will be said of this 'B' hereafter.*

The Psalter of St. Louis, in possession of John Ruskin Esq., is perhaps, for its fine drawing and exquisite finish, equal if not superior to those late XIII century MSS. described. Another choice example is Queen Mary's Psalter, which is considered by Dr. Waagen as in some parts equal to anything done during the XIV century. It is thought to belong to the latter part of the reign of Edward I. It contains a vast number of spirited drawings, illustrative of the popular manners of the age in which they were drawn, and possessing the characteristic English turn for the humourous and ludicrous. It is called by its present name, through having belonged to the daughter of Henry VIII. The embroidered cover is said to have been wrought by her own hands.

We now come to the XIV century period, in which the field is wider, containing many divers styles; some retaining many features of the XIII century, and gradually employing shading more than lines in drapery, yet superb in design and colour, and rich with filagrees and diapers. 17,341, British Museum,† affords one of the finest examples to be met with. It is in the highest state of preservation, and deserves not to be freely handled until the student earns the boon by advanced work.

* Compare this 'B' with the one in Plate VIII., which is of late XIV century, where the drawing is very good, but not so solemn.

† Plate VI. with the music is part of a beautiful page in this MS. The stave consists of only three spaces as in ancient music (sometimes two in others) the four lines being ruled in vermillion for the notes.

In this century was developed the ivy stem and leaf decoration as in the Berri Bible,* and in a lesser sized but
almost equally as good MS. 2897, British Museum, which
contains great variety of treatment—in single line stem
and trefoils, in ivy double line stem and ivy leaves, and
in curling foliage combined with these two ways.† Whole
borders may be repeated in this way, as by bringing
the lower line of ' K ' straight down from the knot end,
and varying alternately the single and double line ivy,
&c., by side of it, sweeping it out similarly on under and
top margins. Other styles almost similar, but employed
in smaller missals, may be seen in 23,145, British
Museum, and Plate VIII. a title-page, from a copiously
illuminated missal, kindly lent by G. E. Street, Esq.,
from his collection.‡ Border of Appendix, is also

The whole painting contains six miniatures—one king playing on bells;
the Jesse below, you have some idea of from Plate V. This illustration
complete, will form one of the sequels for practice with a true copy on
vellum for reference. The backgrounds of figures are gold, not diapered,
surrounded by stems of pale blue and pink, and flowers in scarlet, blue,
green and pink shaded, on grounds of gold, blue, and flushed or clouded
deep carmine-purple. The rod at side is an example of narrow border,
springing at top of a broad miniature with canopy above, and scroll
work below, as in p. 71, somewhat similar in principle to letter ' L,'
(Ruth subject) in Plate IV.

* This is fully described in previous editions; but we may here
merely mention that it is written in double columns and that the miniatures
are square, of the breadth of a column, set as if in gold frames, containing most exquisitely painted subjects, the costumes being richly embroidered, and the diapers of every variety possible, all thoroughly
finished. The capitals commence below these pictures, and sometimes
have springing from the top and bottom of them a straight line border
stem either pink or pale blue, shaded white, with a slight gold band
running with it, breaking out on margins of page into curved stems
with leaves, in some degree like border of appendix, but infinitely finer.
In other instances the capitals stand alone as the ' H ' heading Part V.,
which shows the single and the double line stem.

† These are generally set in a plain field of gold, with scroll work or
diapers sometimes, as in ' K,' p. 60, but this curling foliage becomes lost
in some examples, as if out of harmony.

‡ In this latter, Goliath is streaming with blood and has the stone in
his forehead very prominently shown. The griffins' heads are brought

from the same MS. although it appears dissimilar in general design. One illustration forms an interesting large letter 'D,' from a very fine folio MS., also in possession of Mr. Street.

The culmination of this style may be placed about the date of the battle of Poictiers, 1356, or at least easily remembered by it. Jean, Duc de Berri, who was one of the hostages for the French King after that battle, and lived, as such, in England, was one of its greatest patrons, and the possessor of many valuable and splendid books. The Berri Bible formed part of his collection.

Throughout the xiii and xiv century styles, we find very many curious grotesques such as boar hunts, dogs sitting on leaves, greyhounds chasing rabbits, hares jeering dogs who cannot get at them, crows and birds, coloured in strange manner, such as pink, blue, often beautiful grey, evidently studied from nature, probably in the illuminator's evening walk or pastime—perhaps he went out boar-hunting or hawking occasionally, and so saw how a hound sprang along.

In MS. I. E., IX., British Museum, we have a splendid example of the type of the English Flower School—rosy varied colours, and sparkling with gold.

In regarding the later schools, by many considered the decline, the reader will not, after having his mind impressed with the best xii, xiii and xiv century styles of

in expressively—one looking jeeringly at the giant, as if he had more to do in the affair than being a mere knob to the 'B,' another on bench elbow is wondering at the music of David's harp. The choir-boy, perched on the top of sweeping branch, is chaunting the "Beatus," with the picture of the battle in his mind's eye. If the rabbit does not look sharp, it is all up with him. The monkey may as well give up all his *redoubled* efforts to blow the trumpet. But what has the monkey to do in the piece? Dear reader, the artist had an extensive mind, and not content with representing war in David's time, thought he would draw a parallel scene of a tournament in his own days between two knights, and for fun painted two monkeys, one in fantastic, heraldic costume, to act as heralds—but all is so faint in original, that it is omitted here, but the knights in No. 1 of "Idylls" give a good idea.

illumination, expect good from burdening and confusing his recently acquired knowledge with what becomes inferior in many points as that of the xv, xvi and xvii centuries. Then gaudy and lavish colour with every artifice of fascination was used to atone for the graceful, fanciful design, and solemn earnestness of expression and richness of colour of the previous periods, when it was the aim to enhance and glorify the text as the true end of the art of illumination, instead of as in these later centuries, striving at producing *picture* books more than *text* books—the artists seeking display of their own powers, thus losing sight of their mission. Then began that great change which took place in architecture, in glass-painting and illumination, each from their close connection reacting on the other, and from this and other causes leading to decline.*

* The xiii style of work, from its bold firm drawing inventive power and beautiful colouring, affords without doubt the best instruction even for proficiency for later styles ; but no pupil once convinced of the merits and thoughtfulness displayed in it (as can only be perceived by practice,) will ever delight in any debased styles. The drawing being expressed by lines as in drapery principally, requires less laborious manipulation than xv century and later work or " Renaissance," and if it shows power to give the noblest and best expression by the fewest strokes in a small space, when time is precious, in order to produce many works, the earlier artists have the superiority. If rich imaginative design is the highest of faculties, certainly for this, not only in drawing, but in colouring, the xii, xiii, and xiv century works rank higher than the styles after them, which have less imaginative power, and less design ; for the work is of the realistic and naturalist kind, but very often after all most unnaturally, and unconnectedly, arranged, though almost transcript from nature. All good *decorative* art has from the earliest times been humbly subordinate to nature. This forms a part of what is called " conventionalizing," and thus the xiii century artist was content to employ his imagination in giving us delightful grotesque design out of its suggestions, rather than mock nature by painting grapes growing out of acanthus leaves, or lillies from cabbages. Disobedience! Christ asked, " Do men gather grapes of thorns, or figs of thistles ?" The Renaissance men seemed to say ' Yes.' We will paint any absurdity. We will make you disgusted with violets and primroses by setting pigs, &c., to revel in them—as seen in some examples.

Readers may learn much about the " Renaissance" and its causes

The great feature of the xv century illumination, is
its elaborate perfection of miniature painting—a matter
involving serious labour on the part of a student of this
class of work. It has, however, other distinctive marks
among which are the single line scrolls, composed of a
tendril of ivy, the line in black, the leaves burnished
gold, thickly distributed over all the available portions of
the border.

Sometimes as in the Berri ' Hours' at Paris, in Harl.
2900, and others in our own National library—beautifully
coloured birds are introduced among the tendrils, sometimes a forget-me-not or a columbine, or whatever might
please the artist or his patron.

Jan and Hubert Van Eyck may be termed the ' fathers
of miniature painting,' for though not the originators,

from Mr. Ruskin's "Stones of Venice," particularly third volume,
but he should read the whole work. Third and fourth volumes
of "Modern Painters," also should be carefully read throughout, as much is to be learned useful in illumination, many of
the subjects such as the "Ideal," and "Finish," applying to it as
well as to great artistic work. But we have not finished our
comparison of XIII and XIV century work with later work.
In colour, the early artists obeyed God's favourite decorative
colours as used in the tabernacle—blues, purples, and scarlets
principally, as you will find repeatedly given in Exodus from Chapter
XXVI to XXXIX for the hangings, for the dresses, for embroiderings
and veils ; and the ram's skins, beautiful as they were naturally, were
dyed scarlet for decorations (the goats' and badgers' skins were *not*
dyed as they were used for outer or top coverings, not decoration.) No
browns or such colours for decoration ; we have them in nature—
yes, but seldom in good decoration. It is rarely that brown, or buff,
or grey, or black is used in XII, XIII and XIV century decoration, excepting the two latter, especially black, used only for devils and evil spirits ;
but look at our decoration borders from the XV century, much brown,
grey, and foolish show of gold (not as in the previous ages, using it as a
rare and precious thing that enhances), and black most vigorously.
Another thing, in XV century and later work, much obsceneness was
painted ; the harmless grotesqueness of the XIII and XIV became inde-
cent and some missals delight in cruel scenes and exposing bleeding
carcasses as if painted by butchers—bad signs of the reversed system.
Of course there are exceptions, but generally the above are some of the
distinguishing differences as to *decoration* of early and late MSS.

their improvements placed it on an entirely new foundation. They and their sister, Margaretha, were the joint producers of the great Bedford Missal, another present to Henry VI., whose date Dr. Waagen fixes between 1423 and 1431. They were the first to introduce good landscapes in the place of diapers or bad architectural designs, for their backgrounds. Their studies of natural flowers, and with them insects and realism generally, soon gave a complete turn to the art throughout Western Europe. The same impulse was working among some artists in Italy—whether by their influence, perhaps it would be difficult to decide; but, certainly, they added to the effect by which illumination lost itself in miniature, under such artists as the monk Angelico, Girolamo dei Libri, and his more celebrated pupil, the indefatigable Clovio.

Through this revolution from the decorative to the realistic, we find the study of illumination, as we approach the XVI century, to become exceedingly complicated. It would, if fully followed up, include historical and landscape miniature in body-colour, or 'tempera,' flower, fruit, and animal painting to perfection, and a peculiar method of painting, chiefly Italian, called 'cameo,' or 'camaieu gris.' The ancient 'camaieu gris' is a washy kind of sketching; the later 'cameo' painting an imitation of 'basso-relievo' in monochrome, the lights being white or gold, the shades a deeper tint of the coloured ground, amber, grey, crimson, puce, or whatever it may be.

Among the noted patrons of illumination in the XV century, a prominent place must be assigned to René of Anjou, father-in-law of Henry VI. He was himself an amateur of no mean pretensions. He encouraged, and excelled in various arts, and thus deserves notice—notwithstanding his poor figure in the politics of the day.

A volume that once was his, the work of a Flemish-

c 2

taught French artist, recommends itself for its immense variety of colour.*

It will have been noticed how, by degrees, illumination changed its form from ornamented lettering, to elaborate miniature. It no less decidedly changed its modes of application. From being almost solely appropriated to missals and other devotional books, it became so popular as to be enlisted into the service of poets and novelists.

In the 'Romance of the Rose,' the British Museum contains, perhaps, as glorious a specimen of the kind as ever existed. Another, almost as fine, is the 'Poems of Christine de Pisan ;' both said to be French MSS., though the former undoubtedly reflects the chief excellences of Flemish art.

The first illumination in the 'Romance of the Rose,' has a broad border, with a bloomy, dull, green ground.† The twigs upon it, forming the border, are in lilac—a common colour for buildings, &c., and details generally. Hanging or twining from the stems are, in the right lower corner, carnations; in the upper, columbine; on the left lower side, cornbottle, or thistle ; up the side, forget-me-not ; then daisy, and lastly, sweet-pea in the upper corner. The whole is interspersed with birds and insects in relief. The crimson or scarlet flowers are relieved in the curls with gold.

Next to this MS. in beauty is the 'Hours,' by Memling, who has been characterized as one of the greatest of illuminators. It is in all respects a perfect gem, bound in purple velvet, having a border of open gold filagree, four stones set in silver at the corners, and in the centre of each back, a cameo medallion. The miniature borders have solid grounds of colour, or dead gold. Among the colours are grey, lilac, buff, and turquoise

---

* It is described on pages 45 and 46 of previous editions of this Manual.

† Some initials from this book may be seen in Mr. Shaw's "Alphabets."

green; occasionally crimson and purple, and now and then intense blue. But for the flowers and insects in this MS., we must refer to Dibdin's description of a similar one, the famous ' Hours' of Anne of Brittany in the National Library at Paris.

In the ' Memling' (addit. 17,280) the borders are chiefly on the outer side. In full-page borders, the grounds are cut up into compartments by diagonal lines or twigs laid cross-wise.* For small pocket volumes, an available style similar to the above, but far inferior, is shewn in Harl. 2948, and 2950. The borders are oblong and on the outside of the text, and generally consist of a combination of conventional scrolls, with natural flowers or fruit.†

The Italian style consists similarly of highly ornamented stems and rinceaux, with birds, vases, and flowers upon them. Miniatures are introduced in medallions suspended from the stems, or forming part of the construction of the borders; which are often filled with massive scrolls of acanthus leaves in azure, crimson, and green. These leaves are shaded in a firm and masterly manner, and heightened in the lights with brighter colour or pure white. The designs of Raffaelle in the Loggie of the Vatican are often imitated in Italian MSS.‡

In Spain the art was partly like the Italian, partly tinged with Oriental gorgeousness of taste; for as we have remarked, illumination was cultivated with success and in various styles among the Asiatics. We have now arrived at the very latest period of the art as it was employed in printed books for upwards of a century after the invention which put an end to the business of the *scriptorium*. Initials were often painted into blank

---

* See Figs. 29 and 31 in previous editions of this Manual.

† So many of these examples of late styles have been published, that description or illustration is needless.

‡ A large folio work in British Museum from collection of the late Samuel Rogers contains many fine examples.

spaces left by the printer, sometimes on shields or elegantly shaped cushions, or on medallions of richest colour overwrought as in the MSS., with flowing patterns in deeper colour or gold.

In all later styles, the initials are very similar, the texts vary even less.

Here we close our rapid survey. As, however, so much has been said by various writers concerning the refined finish, and sweet colour of Giulio Clovio, it may gratify curiosity to conclude the chapter with a few observations on his method.

From the examples best known in England, he appears to have profited largely in design by his studies from Giulio Romano, and Michel Angelo, under the former of whom he studied after leaving Girolamo. Hence, as might be expected, he excels also in colour.

His high lights are left untouched, the colour appearing gradually and with caution about the folds and half-lights, and then deepening into the shadows. Everything in features, dress, ornaments, flowers, and other detail, is finished with the nicest attention, not permitting a single slovenly or unmeaning touch. The colours employed in the draperies and trappings, which are disposed with masterly skill, are brown, pale blue, yellow, pink, and gold. His excessive finish is brought about by an incredible amount of stippling and soft gradation of the delicate tints.

Of the time he used to take in his performances, we may have an idea from the current story that on twenty-six miniatures executed in a Breviary for Cardinal Alessandro Farnese, he spent the greater part of nine years; while as to value, the twelve pictures in the " Victories of Charles V," known as the ' Grenville Clovio,' are said to be valued at one hundred guineas each.*

---

* The reader will find a close description of these pictures in Dibdin's Bibliograph. Decam. Vol. I. p. clxxxix.

There is a MS. decorated by his hand in the Soane Museum containing several cameos, and an altar tablet in the South Kensington Museum, also attributed to him. This tablet is of an oblong shape, the texts separated into three columns by lines of ornament; a broad border, with renascimento scrolls and flowers on a gold ground, surrounds the whole. The edges of this border consist of a plain ball moulding in gilt.

Medallions having grounds of scarlet and blue, by turns, are placed at intervals along the border, alternately with coats of arms. At the top of the central column is a painting of the Lord's Supper, and below it two broad panels, scarlet and blue with gold edges and letters of gold upon them. The whole, with its beautiful initials and careful text, forms a very interesting relic of one of the last of the old illuminators.

NOTE :—A most interesting exhibition, by the Society of Antiquaries, of many illuminated MSS., took place lately (June 1861) in Somerset House. Such an exhibition of so many choice MSS. from XI to XVI centuries must have been instructive to all, where the different styles could be distinguished. The cases containing illuminated MS in British Museum, afford much instruction to beginners. If the authorities would show a few more illuminated pages as a series of the different styles, it would be a great boon, from which students, note-book in hand, could be enlightened on the various features. If they would then ascend to the Mediæval room, they would get many useful details from old relics, caskets and enamels, or pick out a good alphabet from the old fresco remains of Westminster Palace—time of Edward III. (XIV century.) The student should also study at the National Gallery the fine paintings of " Fra Giovanni Angelico," and " Orcagna," and those of the Giotto school, which are all about XIV century date.

Part II.

THE MATERIALS.

UNDRY maxims of drawing and colouring, one or two resplendent pigments, and, perhaps, the best method of raising and burnishing our gold, have been lost to us since the XIV century especially. But upon the whole, we are immense gainers in point of the trouble and excessive carefulness necessary for the production of well-ground, softly-working, and brilliant colours; pencils of sweet touch, and, indeed, of every appliance that can shorten our labour; and economise our time. Many old treatises on limning abound with directions for preparations of colours, sizes, and gold —Cennini, for instance, and the Illuminir Buch of Boltzen, and very many MSS. in the British Museum and other collections. Harleian MSS. Nos. 3151, 1279, 1460, 6376. Soane MSS. No. 416 are specimens of the kind. With the exception of scarlet or bright orange, our colours are, for illuminating purposes, everything we could wish, nor could we, as amateurs, hope to surpass the preparations that are supplied with every care for their utmost excellence by the proper makers.

Colours.

I F we include all that is called illumination from the earliest times until the XVI century, a small stock of colours will scarcely suffice. Of those manufactured at the present day, the following may be used.

YELLOWS.

Cadmium Yellow
Gamboge
Lemon Yellow
Mars Yellow
Naples Yellow
Raw Sienna
Yellow Ochre

REDS.

Brown Madder
Carmine
Crimson Lake
Indian Red
Light Red
Orange Vermilion
Scarlet Vermilion
Rose Madder
Rubens' Madder
Vermilion

BLUES.

Cobalt
French Blue
Intense Blue
Indigo
Smalt
Ultramarine Ash

ORANGES.

Burnt Roman Ochre
Burnt Sienna
Mars Orange
Neutral Orange

PURPLES.

Burnt Carmine
Indian Purple
Purple Lake
Purple Madder
Violet Carmine

GREENS.

Emerald Green
Oxide of Chromium
Olive Green

BROWNS.

Burnt Umber
Sepia
Vandyke Brown

BLACKS.

Ivory Black
Lamp Black

WHITE.

Chinese White

From these we would select and recommend the following five lists, the first being the fewest number of colours you can possibly work with, viz.:

1st List.—Gamboge, Cadmium Yellow, Crimson Lake, Vermilion, Cobalt, French Blue, Emerald Green, Lamp Black, Chinese White.

2nd List.—Lemon Yellow, Gamboge, Cadmium Yellow, Rose Madder, Crimson Lake, Vermilion, Cobalt, French Blue, Burnt Sienna, Emerald Green, Vandyke Brown, Lamp Black, Chinese White.

3rd List.—Lemon Yellow, Gamboge, Cadmium Yellow, Rose Madder, Crimson Lake, Carmine, Orange Vermilion, Vermilion, Cobalt, French Blue, Burnt Madder, Emerald Green, Green Oxide of Chromium, Vandyke Brown, Lamp Black, Chinese White.

4th List.—Lemon Yellow, Gamboge, Cadmium Yellow, Mars Yellow, Rose Madder, Crimson Lake, Carmine, Orange Vermilion, Vermilion, Indian Red, Brown Madder, Cobalt, French Blue, Neutral Orange, Burnt Sienna, Burnt Carmine, Indian Purple, Emerald Green, Green Oxide of Chromium, Vandyke Brown, Lamp Black, Chinese White.*

5th List.—Lemon Yellow, Gamboge, Naples Yellow, Cadmium Yellow, Mars Yellow, Rose Madder, Rubens' Madder, Crimson Lake, Carmine, Orange Vermilion, Vermilion, Indian Red, Cobalt, French Blue, Smalt, Mars Orange, Burnt Sienna, Purple Madder, Burnt Carmine, Indian Purple, Emerald Green, Green Oxide of Chromium, Vandyke Brown, Lamp Black, Chinese White.

Our space will not permit us to enter into the peculiarities and properties of these colours, which, indeed, we believe to be generally known, from there being several works and treatises in which they are fully discussed.† We may, however, draw attention to the permanent character of the colours we recommend, and would warn our readers against using Chrome Yellows, Red Lead, and Pure Scarlet as not being lasting. Pure Scarlet is fugitive and the others in time turn black.‡

It only remains under this head for us to select the form of colour, or mode of preparation most suitable for illumination, a matter which we will now proceed to determine.

In some sense, the best form of colour may be said to be that which the Artist likes best to work with, and with

---

* We observe it mentioned in Mr. Digby Wyatt's work, that Messrs. Winsor and Newton have placed these four lists into boxes (complete with colours and materials) of the respective retail values of £1 1s., £1 11s. 6d., £2 2s., and £3 3s. They have adopted our fifth list, and have placed the colours into a box, which, (being completely fitted with other materials) is of the retail value of £5 5s.

† Vide Field's Chromatography, Winsor and Newton—Digby Wyatt's Art of Illumination, Day and Son—Rowbotham's Art of Landscape Painting in Water Colours, Winsor and Newton, &c.

‡ Vide Mr. Digby Wyatt's Art of Illumination, page 83.

the manipulation of which he is best acquainted. The old illuminators used powder colours, mixed as wanted with white of egg, and curious preparations of colour-cloth. In these days each artist was his own artists' colourman, and having ground up his crude colours into powder, kept them by him for mixture, or always ready in bottles. Availing ourselves of modern preparations, we may pronounce for *Water Colours* as being best adapted for illumination.

There are two kinds of Water Colours, viz., *Dry Cake Colours* and *Moist Colours*—and of these two the latter are preferable as giving out the greater volume of colour, and possessing the greater tenacity or power of adhering to the surface of the material on which they are used. Of Moist Colours there are two descriptions, viz., *solid* and *liquid*. Some of the colours will not keep well in the latter form, and moreover there is a waste in using them when only moderate quantities are required, as the colour cannot be replaced in the tube when once squeezed out. Hence it is best to use the *solid* Moist Colours.

The ordinary kinds of solid Moist Colours are those placed in pans, and those made up *per se* in wafers. They are both apt to get dusty and dirty when once uncovered, and therefore cannot be said to give complete satisfaction for illuminating purposes, wherein it is indispensable that the colours should be kept very pure and clean. Messrs. Winsor and Newton have got over the difficulty by placing their Illuminating Colours (preparations of solid moist character) in small china galli-pots, *with lids*, and have consequently turned out the best form of illuminating colour with which we are acquainted.*

---

* Only lately, still greater improvement has been made on these gallipots, by making them of glass, so that *the colours can be seen through the lids.*

It may be as well to mention, that under the head of 'colouring,' will be found a full list of mixed tints and compounds, made from the colours forming the foregoing lists.

Inks.

E will now briefly touch upon the inks used in illuminating. They are Liquid Indian Ink, (Liquid Lamp Black, Liquid Prout's Brown,) and Liquid Coloured Inks. Of these, the Liquid Indian Ink is most useful, superseding both Liquid Lamp Black,* and Liquid Prout's Brown. The coloured inks are only occasionally required.

If the Indian Ink or Lamp Black be rubbed down from the cake, it must be with very weak gum water, to prevent the ink from spreading unmanageably from the pen, and, moreover, to give it a brilliancy similar to that of the inks used in the MSS.

For coloured inks such as scarlet, azure, or purple, more gum, and a somewhat thicker mixture is made. Scarlet may be made with pure vermilion, or vermilion and a little good red-lead—the gum will assist the permanency of the colour. Azure, with smalt or ultramarine, paled, if necessary with white. And similarly other colours, for the mixtures of which directions will be given hereafter.

The thickness of the ink must depend on the purpose it is intended for. The text may require it to be somewhat flowing, though not too thin, while outline drawing, or rather finishing, will require it to be as thick as can possibly be worked.

---

† The solid moist form is best for Lamp Black, and we have consequently placed the colour in our lists (vide page 26.)

Vellum, Boards, and Papers.

ELLUM can be obtained ready prepared for the illuminator, in pieces, whole skins, or ready made up into block books. It is an admirable material to work upon, but is somewhat expensive.

It must not be supposed, however, that vellum is the necessary or only material for our own purpose. Paper of various kinds was in general use for MSS. in the East, long before its employment for that purpose in Europe. Perhaps two thirds of the MSS. discovered in various Greek and Syrian Monasteries, chiefly portions of patristic theology, are written upon paper. Vellum certainly has the better chance with the lapse of years, though by no means the easier to work upon. Before the days of Bristol and London board, attempts were made at producing a kind of strong card, by pasting the leaves of a book by three and four together before writing upon them. A book of this very kind, and fairly illuminated, occurs among the Harleian MSS. No. 6103. (Harl. 3314, and 3349, &c., are illuminated on paper.)

Bristol and London boards, however, are now (by the aid of machinery) beautifully made, and quite supersede anything that we can make with our own hands.

Paper has good precedent for its use in illumination, nor is it improbable that if paper such as is now made had been accessible in former times, very many more 'Codices Chartacei' would have been numbered among illuminated MSS. Good drawing paper, with a fine, firm grain, and having an ivory-like surface, but without gloss, is, with the single exception of durability, very nearly as good as vellum. We believe that there are papers specially manufactured for the illuminator.

Pens.

STEEL or iron pens were known centuries ago ; but as these are apt to tear up the surface of the paper, we should have at hand several good quills to print with. The reed was used for pens in former days. For texts the student should try either steel or quill pens till he finds one suitable. Some may be cut straight across, or at right angles to the axis of the pen ; others, as may be found convenient, somewhat slanted in the cut, as for writing Greek. Oriental writers, at least, among the Hebrews and Arabians, hold the reed as nearly perpendicularly as possible. A broad, almost unyielding . point will give a backward stroke as clearly as a forward one without taking it from the paper, as may be seen in the beautiful Hebrew and Persian character. Careful as were the Orientals in their writing, they were not more careful than the calligraphists of Western Europe from the xii to the xvi century.

For finer writing, the pen should be cut with a much longer slope in the nib. Such a pen should be made from a strong quill, or it will not bear the requisite pressure of the hand. A steel pen, from its elasticity and the need of nicely regulated pressure, is too uncertain, at least, until some dexterity of stroke has been attained.

For fine italics, however, a smoothly finished long-nibbed steel pen may be found exactly suitable ; while for fine drawing, such as is continually required in the best work, that of the close of the xiii century, one of Gillott's architectural pens is perfect.

Whatever may have been the dimensions of the pen or brush with which the artist accomplished the fine white lines on the initial and diaper grounds of MS. Harl. 4381, it is unquestionable that they were drawn by a most fearlessly skilful hand. We can have no idea of their fineness from a mere engraving nor unless we have seen something of the kind in original work.

Brushes.

EW brushes are required for illumination, but from the peculiar character of the work, and the nature of the colours, &c., employed, it is requisite that the right kinds should be carefully selected.

For general use, the red sable brushes in goose, duck, and crow quills should be employed; say one goose, (for large grounds), two duck (for ordinary work), and three crow (for fine linings on initials, &c). The red sable is preferable to the brown sable, or other hair, as being stronger and firmer at the point.

An ordinary flat camel's hair brush will suffice for damping the back of gold paper, washing over weak solutions of gum water or ox-gall, &c., &c.

When gold leaf is used, a soft camel's hair brush (of swan quill size) may be found useful for touching, smoothing, &c. For laying down the gold leaf, a very thin flat brush is required, called a ' gilder's tip.'

Burnisher and Tracer.

DOG'S tooth set in a stick was the ancient burnisher, for general purposes; but agate has now been long in use. Some illuminators prefer and use the ' claw' shape, but that of the ' pencil' may be said to be the best, from the facility it offers to beginners, in allowing a fair sweep of hand over the gold. A long sharp pointed ivory tracer, is requisite for tracing and useful for indenting gold diapers.

Metallic Preparations.

GOLD Leaf, Gold Paper, Shell Gold, Saucer Gold, Gold Powder, Gold Paint, Gold Ink, Silver Leaf, Shell Silver, Saucer Silver, Silver Powder, Silver Ink, Shell Aluminium, and Shell Platina, are the various forms of metallic preparations used in illumination.

Their mode of application will be found explained in our chapter on ' Gilding.'

Besides colours, materials to work upon, pens, brushes, burnisher, tracer, and metallic preparations, there will be required an eraser, compass, rule, pencil, Indian rubber, sponge, cotton wool, some tracing paper, a small **T** square, and other small sundries.

A bottle of gum water will be necessary, and also one of liquid ox-gall. A little of the former, mingled with water, is used to impart brilliancy to colours and for other purposes mentioned in this book. The latter (likewise diluted) will be found useful for removing greased particles from the surface of vellum, paper, &c.

It will also be requisite to procure some Water Mat Gold Size, and Raising Preparation—of these we shall treat in our chapter on ' Gilding.'

Part III.

~~~~~~~

HOW TO SET TO WORK.

NDER the heading of " Outline " in former editions, was discussed the manner of proceeding to work, a subject we will now take up under a fresh title. Here we begin to learn the immense distinction and difficulty there ever is between imagining and doing.

During the moments which we have occupied in reflections upon the history and principles of the Art, doubtless many entrancing visions have presented themselves of subjects resplendent with gold and colour; and grand determinations have been formed for triumphs in the glorious pages with which we would astonish and delight the eyes of our friends ; and ourselves become rich in precious books, all our own handy-work.* Let us now put these schemes into execution. Illumination is truly a splendid art, but it cannot be attained at once. We must submit to a somewhat tedious

* It should be carefully considered that a laborious and careful pur-

D

application to study and incessant practice, and be fore-armed against the little disappointments which naturally will occur. After hours or days of happy progress, let us not permit an unfortunate dash of colour, or an obstinate piece of gold to dishearten us from beginning afresh. It is a lesson of caution that, no doubt, we have been in need of.

If, after careful reflection, we still err, then let us trust to practice also, and most assuredly our work will soon become easy to our hands, and the bright visions of our imagination be fashioned into permanent reality. Let us endeavour to make up our minds to complete to the best of our ability everything we commence, and to avoid falling into the habit of throwing away one piece after another unfinished.

First, then, let us have a perfectly definite idea of what we have to do, it being presumed that we have gained some power of eye-judgment and skill of hand in drawing. It is common to suppose that if we could only manage the pencil we should find no drawing difficult. In some degree, this is true, for facility of hand is only acquired by practice. But the great difficulty is in the seeing, not the handling. We do not imitate perfectly, because we have not seen all that there is to imitate, and however forcibly this may apply to the distinctness and reality of form, it applies much more forcibly to the subtlety and changeful mystery of colour. Correctness of

suit such as illumination, should not be thrown away on frivolous and useless subjects. The art was meant for books and important documents, and surely illuminators can be at no loss. Try small missals; for instance, of Songs and Prayers of Women of Scripture; or modern poetry, such as Tennyson's immortal poem of the "May Queen," with varied ivy borders, and miniatures of principal scenes; or his "Sea Dreams;" or a set of choice poems such as Kingsley's "Three Fishers," "The Border Land," "The Burial of Moses," &c. The art may also be well employed on important documents, such as title-deeds and settlements, church contracts and endowments, or marriage contracts, family registers, or public addresses to illustrious personages, and titles of architects' mediæval designs.

eye will very soon produce correctness of hand. Our first acquirement then, must be, to learn how to observe. If we can do some one thing with accuracy, although not rapidly, we have made better progress than if we had gained a bold style of throwing off rapid but flippant inanities of gold and colour. We must commence our drawing carefully at first, even although slowly, for it is gain in the end, as thereby we shall become accustomed to ensure correctness afterwards. If we commence wrong, all will be wrong to the end, in false quantities of colour, and cause no small disappointment.

It is often remarked that the source of beauty is in changefulness. Hence the preference of the curve to the right line—of the curve of many centres to the curve of one—and of that whose centres change with subtle graduations from side to side—to that in which they range along the same.

It is for this eminent characteristic of beauty, combined with splendid colour, that we would commence our studies with the illumination of the xiii and xiv centuries.

In commencing our work (if for a book), we must take a comprehensive conception of how we will treat the illuminating throughout, from title-page to end. If it is to be a subject occupying some fifty or hundred pages, we must calculate what labour we will bestow upon it, and what regular hours we will devote to it. If a worthy subject has been selected and is intended to be very rich, then we may perhaps wish complete surrounding borders on every page, as in different illustrations shown; or less work, as in Appendix border; or less still—half of it, perhaps, joined in with a capital at side, or a slim border, at left margin, running down from a large capital at top as P in Plate III. or if our capital is in middle of page with a slim border, up and down from it, such as in Plate V., and B Part IV. At the same time we should try to combine the force or general expression of

these pages effectively in our title-page, making it the
best adorned and chief design of the following ones, thus
keeping all in harmony, as in the title-page medallion-letter
'I,' Plate IV., which commences a Bible, the features of
which exist in the most perfect consistency, but with
great powers of variety to the end of Revelations, and even
to the long register of every name in Scripture. We must
take care not to mix the style of one century with another;
by and by we may discriminate so far as not to introduce a feature of late xiv with late xiii century work.

If the subject is only one page, such as a few verses of
poetry, or a presentation address, we should try and give
some expression or design connected with the subject in
our border, or miniature or capital. Armorial bearings*
may be required perhaps, which we may see well hung in
F. S. A's "Idylls of the Kings."

After carefully considering, or turning over in our mind,
the arrangement of our piece of work, we may proceed to

Ruling for Margins and Lettering.

F the reader looks at old MSS. he may
see pierced marks for ruling. This was
done through many sheets at once, and
then ruled throughout with very faint
light red or light blue margin, and
parallel lines for border and text, without
regard in the meantime to situation of capital. It
would be advisable for the pupil to follow this method, as in Plate VII., but if positions of miniatures
and great initials are fixed, do not rule through
them. The lines which may run through a capital will
be covered or washed out by the colour. The lettering
was written within these ruled lines, without any other
pencil or guiding lines, as will be explained hereafter.

---

* Messrs. Winsor and Newton will shortly issue a concise work on
Heraldry, which will save much searching and trouble to artists.

Border.

ITHOUT the confines of text and margin lines is generally placed the border. So many examples of borders are given that they do not require much explanation. The principal thing to be observed is, to have them in harmony with the miniatures and capitals, with which they should be connected in some way. Our first capital may form a border itself. Plate IV. for instance, (late XIII,) with the four quatrefoils, (left for the reader to fill in for practice), is only half the great initial 'I' commencing Bible at Advocates' Library, Edinburgh, containing the Seven Days of Creation, and then a square miniature of the Crucifixion, with a slighter border on other three sides. The quatrefoils have diapered backgrounds like Ruth letter in centre, of red-purple, and smalt alternately, the outer grounds varying with them in plain blue against purple, or purple against blue; the stems are pink or pale cobalt, as are also the bodies of griffins, and the leaves are pale pink, scarlet, and green set in gold. The slight border is coloured on the same principle, and is connected with main border by birds and hunting scenes, and altogether forms a grand page.* Plate V. is letter 'L' of 'Liber Generationis,' commencing St. Matthew's Gospel,† forming border at top, but with desire to have a border all way down, and so a thin one is painted as in Plate VI.; in which latter plate, the twisted branching with miniatures gives a suggestive example of beautiful bordering. Note the variety even in an apparently uniform border, as the patterns thereon in Plates I., II., &c. There are other styles of bordering seen in Plates VII. and VIII., Appendix, and Plate IX., also in Letters 'H,' Part V., and

* The whole page (4to imperial,) will form a separate sheet for practice.
† Forms one of sequel complete.

'K,' p. 60, parts of which explain the varied work that may be repeated all round a margin.

In the later styles from xv century, the borders become broader, and the page is almost all border, with pieces of flowers, birds, and creeping things, and the miniature inside, with a small capital below, and but two or three lines of text. These borders are sometimes decorated with imitations of mounted jewellery, as in Harl. 3045. Other borders such as Italian, are designed with vases, flowers and birds, called "Raffaelesque" work, but all these later styles take immense time to finish.

Initials.

OR the design and drawing of initials or main capitals, we can only refer readers to the examples throughout the book and sequel, and certainly it will prove hard if they cannot get from one and another, something to suit the most fastidious taste and arrangement. The smaller capitals, heading paragraphs throughout Manual may be employed for beginnings of sentences, but without border, and should be done in solid blue, with curling work round in vermilion, or solid vermilion with blue curling, or raised gold with either blue or vermilion surroundings as in 'M,' Plate V., or 'A' in opposite page, but not to be outlined in black. When forming small leading capitals, they may be in raised gold, on the solid square grounds of smalt or purple, lined black, finished with white hair vein stems and flowerets. Their colouring and that of borders, is treated of in the proper place, in next Part.*

---

* After carefully sketching in our border, miniature, or grand first initial, we would do well to cover these parts up with a margin or mount of thin white paper, cutting the inside of it so as to form a flap over the text we are about to commence, and which will take long. We thereby save rubbing or greasing the border, and have a cover for the whole surface, when put aside.

Text or Lettering.

HAVING sketched in a tolerably fair piece of work, it is to be hoped that we will give our mind to mastering the lettering or text. This is the main object, the border or capital being only meant to beautify or adorn the words we think worthy of being patiently and thoughtfully inscribed. If done carelessly, it will only show that we attach little importance to perhaps a most sacred and choice passage. If we do it carefully at first, the habit will become facile and easy. If the reader has never attempted an old English alphabet before, so much the better, unless it was properly done; for there is no habit so ill to conquer as a bad hand, with words tumbling here and there; and y like g, or l like t, or w like m, so that they are painful to read.

The pupil should see some good MSS. of xiii or xiv century date, and copy out alphabets on paper, faintly ruled in one-eighth of an inch parallel lines, also in quarter of an inch, three-eighths of an inch, and half inch lines, endeavouring to form every letter perpendicularly to the line, and of a uniform thickness, noting all the little turns, as in alphabets here given from good MSS. which are written between the ruled lines, and which he should attempt without guiding lines, after his first alphabets on the lines.

abcdefghijklmnopqrstuvwxyz;:.
abcdefghijklmnopqrstuvwxyz.-

| commandment | I. III. IV. |
| commandmt | XXVII. &c. |

Ancient MSS. were generally written in glossy black ink, which can be made by lamp black, gum water, and Indian ink, well mixed and rubbed together in a saucer until quite thick. Some parts, such as short introductions or prefaces to chapters are written in vermilion, other parts in blue. Some fine missals are written in raised gold and fine blue, such as the beautiful missal of Yoland of Navarre, in possession of Mr. Ruskin. The gold lettering is a difficult process, as underneath is a composition, over which the gold is laid, thus giving double work. The blue may be cobalt, with a little smalt and Chinese white mixed thick so as to be raised like the gold.

When numbers occur, they look well if done as above, the upper and under dash strokes being red, and body of number between in blue, or visa versa.

Having finished in all our black text, we may now cover it up, and uncover round our border, to prepare for colouring.

Part IV.

~~~~~~~~

COLOURING.

ESIDES the colours already mentioned in "Materials," several others are requisite, that are only to be obtained by mixing on the slab or palette, such as broken hues, employed in backgrounds, and shading for the more brilliant colours and tints in the ornamentation.

We cannot here enter upon the study of principles and maxims of colouring, so well explained in other works,* but we may give some practical directions for laying on such colours as are named.

First, then, everything connected with

* Vide "The Principles of Colouring in Painting," by Charles Martel—Winsor and Newton.

the painting should be scrupulously clean and free from dust. Distilled water should be used, or, at least, soft water that is perfectly clear. A very little gum-water, in some cases, and a little Chinese white, more or less depending on the colour it is to be mixed with, should be added to the colour as it is mixed. The sable pencils should be in readiness, two or three, or more, according to the work, and habits of the operator.

Perhaps this is the best place to notice the manner of using the Chinese White. On being taken from the bottle, it is found to be exceedingly viscid, and troublesome to work, clogging the point of the pencil. Of course it should be diluted with pure distilled water, but as this renders it too thin for the firm, fine lines and dots so often wanted, it must be left a few moments to evaporate and thicken; if still viscid, it should be thinned again and left. When thus put out of the bottle or tube and thinned, it will be found to be even better for working a day or two afterwards than at first. All that is required, as it will be dry, is to dip the pencil in water before working it upon the white, and make a good point before transferring it to the illumination. Unless these precautions be observed, the use of white will be attended with continual vexation.

When a compound colour is required, sufficient for the work in hand should be mixed up at one time, lest, more being required, the second tint differ from the first, when a disagreeable 'patchy' appearance will ensue. Compound tints should be kept extremely clean in tone, muddy, or dirty tints being fatal to that exquisite purity of colour for which illumination is so famed. It is a great mistake to suppose that dark tints are necessarily somewhat dirty; on the contrary, they can be kept as clean and clear in tone as the most vivid combinations.

We will now proceed to furnish a list of colours and mixed tints, and, in so doing, will drop all technical phraseology as much as possible, so as to convey the information clearly, even to the tyro.

Table of Colours and Mixed Tints.

YELLOWS.

Vivid high-toned Yellow, or Primrose.—Lemon Yellow, Lemon Yellow and White, Gamboge and White.
Bright transparent Yellow.—Gamboge.
Rich glowing Yellow.—Cadmium Yellow.
Clear, transparent Yellow.—Mars Yellow, Lemon Yellow and Cadmium Yellow, Lemon Yellow and Gamboge, Gamboge and Mars Yellow.
Rich brown Yellow.—Cadmium Yellow and little Purple Madder, Cadmium Yellow and little Indian Red.
Buff Yellow.—Cadmium and touch of Burnt Carmine, Orange and little White.

REDS.

Vivid high-toned Red.—Orange Vermilion.
Deep opaque Red.—Vermilion.
Bright transparent Pink.—Rose Madder, Rose Madder and touch of Carmine.
Opaque Pink.—White and little Orange Vermilion, White and little Vermilion, White and little Indian Red, White and touch of Carmine, White and little Rose Madder.
Rich glowing Crimson.—Crimson Lake, Carmine.
Chocolate Red.—Vandyke Brown and Carmine, Vandyke Brown and Crimson Lake, Burnt Carmine and Orange Vermilion.
Russet Red.—Carmine and Indian Red.

BLUES.

Bright azure Blue.—Cobalt, Cobalt and White.
Rich strong Blue.—French Blue.
Brilliant purple Blue.—Smalt.
Deep dense Blue.—French Blue and little Black.

ORANGE.

Clean pure Yellow Orange.—Mars Orange, Neutral Orange.
Deeper Yellow Orange.—Burnt Sienna.
Intensely brilliant transparent Red Orange.—Carmine over a ground of Gamboge.
Rich glowing Warm Orange.—Cadmium Yellow and Carmine, Cadmium Yellow and Orange Vermilion, Orange Vermilion and little Lemon Yellow.

PURPLES.

Rich cold Purple (Violet, Lavender, &c.).—Indian Purple, Indian Purple and French Blue, Cobalt and little Rose Madder, Cobalt and little Crimson Lake, Cobalt and little Purple Madder, French Blue White and little Rose Madder, French Blue and little Crimson Lake, French Blue and little Burnt Carmine.

Rich warm Purple (Puce, Marrone, &c.).—Purple Madder, Burnt Carmine, Crimson Lake and little French Blue, French Blue and Carmine, Rose Madder and little French Blue, Rose Madder and little Cobalt, Crimson Lake and Cobalt, Burnt Carmine and little French Blue. White may be added with any of these.

Greyish Lilac.—Cobalt and Brown Madder.

GREENS.

Vivid high-toned Green.—Emerald Green, Emerald Green and Lemon Yellow.

Bright Apple Green.—Emerald Green and little Oxide of Chromium, Emerald Green little Oxide of Chromium and little Lemon Yellow, Lemon Yellow and little Cobalt.

High-toned transparent Green.—Gamboge and little Cobalt, Cadmium and little Cobalt, Gamboge and little French Blue, Cadmium and little French Blue.

Low-toned transparent Green.— Cadmium Yellow, French Blue and very little Crimson Lake, Lemon Yellow, Cobalt and very little Rose Madder, Cobalt and little Gamboge, Cobalt and little Cadmium Yellow, French Blue and little Gamboge, French Blue and little Cadmium.

Light Opaque Green.—Oxide of Chromium and White.

Deep Opaque Green.—Oxide of Chromium.

BROWNS.

Pure Brown.—Vandyke Brown.

Rich warm Brown.—Vandyke Brown and little Burnt Carmine, or Crimson Lake, Purple Madder and touch of Cadmium Yellow, Vandyke Brown and Brown Madder.

Cold Brown.—Vandyke Brown and Indian Purple.

Yellow Brown.—Indian Red and little Cadmium.

Stone Drab.—Vandyke Brown and White, Yellow Ochre and White.

BLACK.

Dense Black.—Lamp Black.

WHITE.

Pure White.—Chinese White.

GREYS AND NEUTRALS.

Grey.—Black and White.
Pearly Grey.—Black and White and little Cobalt.
Slate Grey.—Black and White and little Crimson Lake, Black and White Indian Red and Cobalt.
Silvery Grey.—Black and White, Cobalt, and Rose Madder.
Clear warm Neutrals for shading.—Orange Vermilion and Cobalt, in various proportions.

Various proportions of colour may be tried, particularly for the greys, neutrals, and quiet compounds, and the most pleasing and suitable, should be carefully noted for use.

Brilliancy is obtained by contrast—setting out a light colour on a dark, and also by gradation. Suppose a scarlet over-curling leaf, for example. The whole should be painted in pure orange, with the gentlest possible after-touch of vermilion towards the corner under the curl. When dry, a firm line—not wash—of carmine, passed within the outline on the shade side only of the leaf, will give to the whole the look of a bright scarlet surface, but with an indescribable, superadded charm, that no merely flat colour can possess. If a scarlet berry; pure orange as before, for the first painting, while still rather damp, drop into this, near but not close to, the edge furthest from the light, the smallest possible drop of vermilion. When quite dry, finish with a minute globule of white, just where the light is supposed to fall, and the berry will appear glossy.

There are, however, positions in which flat colour is absolutely necessary, and it should be not only flat, but dead, that is, entirely destitute of gloss. In painting grounds, whether of diapers, or initial, or for the flowers of later style, all stipple, flocciness, or mottling, should be sedulously avoided. The colour should be moderately thick, yet not viscid, and the passes with the brush made deliberately in one and the same direction, so as to give

to the finished surface the appearance of enamel. Not a
wave nor spot should be visible. And even in miniatures
or pictures of any kind accompanying the illumination,
the work looks far better and more consistent with that
bloomy 'mat' surface, than with all the sparkle, or
intensity of depth, afforded by glaze and varnish.

Two of the characteristics of the XIII and XIV styles
with which we commence, are :—I. That it is purely
surface decoration, there are no thrown shadows, no attempt at imitating solidity or relief in the object selected
for ornament. The colours are pure and well-chosen, merely
gradated from deep to pale, and sometimes up to pure
white.   II. The drawing is then done throughout with a
firm, black, glossy outline which gives it brilliancy, and is
further heightened by a thin white hair line edging the
black, as seen in all the illustrations, (although hair
line is shewn black in them, but white can be painted
over after ground colour is laid on).   Sometimes this
black outline is rather thick, and sometimes fine and
delicate, evincing the utmost skill in the management of
the pen.   Before this lining in, our colours may look
dull and hopeless, but the moment we begin to work in
the black line, a pleasing effect commences, still more
gratifying after the white edging and flowerets are painted
—when it is then finished.

In colouring, we should only mix up two or three
of the colours that will be most required ; as for blue
and purple grounds, and an enamel pink and pale
blue, and a little scarlet, for which have separate
brushes; and as we proceed, say first with blue ground,
we must consider that a contrasting colour, such as a
pink stem is to be near it, which may be continued pale
blue, and then a purple ground will be required ; then
we must not have two scarlets near each other, nor leave
too many white leaves or buds.   This is best explained
by looking at some illuminations.   We must for the meantime get out of our head all other laws and rules for

colouring, and learn the manner in which the mediæval
men coloured; who we will find had certain broad
principles, but we shall find also that they were not
fettered by them, but for variety and to please the eye
despised all formula.

We must of course colour blue parts with blue, and
purple with purple, &c.; not any different colour over
another; and all colouring should be mixed up of the
sufficient tone and body to be laid on once, and not in
two coats. We should slightly damp our paper or vellum
previously (especially for blue parts) by a gentle washing
with a soft broad camel hair brush and water, and in a few
minutes lay on colour. The general colours of grounds,
of borders, and capitals of xii century, are so un-
limited in variety as to defy description almost; but
in xiii or xiv century MSS. they are principally blue, red-
purple, of different tones, and gold; the latter used
thoughtfully and sparingly, where it would best tell, and
yet in some cases nobly displayed completely on whole
ground, round God, or saints, as Plates I. and II., or great
subjects; such are often diapered with an indented pattern
in xiv century work. When a mass of blue ground occurs
round a figure, or in a blank space, it is diapered with
black lines, see F, Plate II.; and Plate V. if purple, with
scarlet lake lines, &c., as will be explained under "Diapers."

The ornamental stems, and curling branches, dragons,
&c., are generally of a glossy enamel pink, or pale blue,
like cobalt and white; the former colour commonly standing out on a deep smalt ground, and the latter on a deep
purple ground. Leaves and buds at ends of stems are
coloured scarlet, blue, pink, white, and green, like
emerald and oxide of chromium mixed; pale buds are
sometimes of a yellowish tinge. In some xiii century
work, little green was used. All these stems, dragons,
&c., are gradated by a darker shade, the dark side has
small white hair circles and dots painted thereon, as A.,
Plate II., &c., Leaves are also gradated, or heightened

with white. Thin stems have generally only a white line
on one side to show gradation. Initials or main capitals
with their details are treated and gradated in the same
way; being either pink or blue; if with a bar as in A,
or E, or F, it is painted scarlet for variety, and gradated.

Gold is only used on grounds, nimbi, and rich dresses,
border lines, immediately round leaves, also for balls
attached to buds, &c., or interspersed on grounds, and for
important text, or smaller capitals—never for stems or
leaves, except ivy leaves in xiv century.

The ivy borders which belong to the xiv century
style have no ground colours. Their stems are invariably either pale lilac or pink, or pale blue, with scarlet
knots from which the branches spring, or else are of a
single line, as in H, of Part V. The leaves are painted in
raised gold, in blue, in scarlet, in pink or lilac; all
stems, and leaves (excepting gold), being gradated with
white, and then finally drawn in with black, leaving the
points sharp. Some more information will be given on
colouring in a concluding part of this work.

Of a transitional style, we would mention a French
MS. Egerton 1070 B. M. for the infinitely changeful
character of its colouring.*

The rods used in the simplest borders, which pass
merely up the left side of the column of text, (of which
there are sometimes two in a page) are of burnished gold,
drawn in outline with pen and ink. The foliage is of
two colours to each tendril. It is painted in almost
flat colour, deepened in some cases on one edge of a
leaf. Sometimes a kind of core, pear-shaped, is placed
in the bosom of the leaf, and softly shaded, and if
there be no difference of depth on the edge, the colour
deepens under the curl, as if the ornament were placed in
a strong, but somewhat diffused, light. The alternate
colours may be used on the right and left of a single leaf,

---

* See farther description of this MS. in former editions.

still not interfering with the back and front which may vary as before. Thus: the side *a* might be lemon; *b*, pink; *c*, scarlet; *d*, blue; *e*, lemon; *f*, pink, or merely light on one side, dark on the other. Of the rod-ornament, the whole length generally holds about seven of these tendril-like leaves. When the rod takes a serpentine form like a stem, it carries flowers, and birds, and butterflies, particularly in the eye of each spiral, of endless variety in shape and brilliancy in colour. If going quite round the page, one quarter has one pair or set of colours, another another, a third like the first, and fourth like the second as in heraldic quarterings.

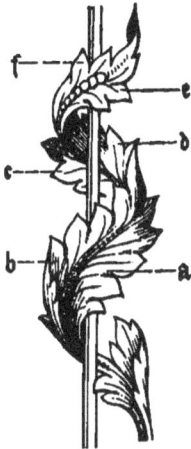

In Spanish and Italian borders and initials much beauty of colour, in some cases of the highest class, is to be found. We refer for example to B. M. Add. 15,813 and 21,120. The latter is curious for the forms of the great initials.

For the Flemish style of Add. 17,280, the grey is composed of black, white, and a little red of some kind; the blue is French blue and white, the red is vermilion and crimson lake.

A stone-drab ground has the thistle or vine leaves the same colour; finely heightened and outlined with gold, and shaded with rich brown. Deep cast shadows are on the gold. These twigs and leaves may be any colour, pea-green, grape violet, &c., modified by white to a bloomy tint. Emerald green leaves are shaded with grey, not green or brown; but other greens have all these shadings. Blue also has grey shading. Red has rich russet.

Initials are all of grey lilac on burnished gold grounds. Ground colours vary as in preceding style.

We do not condemn scientific formulas, but in the present state of the *science* of colour, it is rash to say

E

what is and what is not right, by law. In delicate harmonies, considerable licence must be allowed to what is called taste. Whatever the numerical formulas may do in preventing us from utterly disgusting ourselves, we may depend that we shall only attain to excellence in colouring by careful study of beautiful examples. Reynolds's maxim of constant copying was practised ages before by the Italian masters ; and it is so still, as the hundreds and thousands of studies left by deceased artists testify. Twenty good colour studies, patiently copied, are worth all the formulas in the world.

Diapers.

CROSSED patterns or diapers have been employed from the earliest—even Egyptian times—as important decorative features in buildings, in woven cloths for hangings, and for dresses; and during the middle ages were extensively used for similar objects, but for illumination and glass-painting principally. It was asserted in note, p. 2, that a successive obedience to certain broad rules was transmitted from one generation of artists to another, down to the XIV century. Many of the following diapers, p. 51, in XIII and XIV century MSS., exist in Egyptian sculptures, painted frescos, and metal work, &c., as Nos. I, II, III and IV. No. V. is the actual diapered brick pattern of façade of Ducal Palace, Venice, a type exactly similar exists on a fresco from Thebes, and the same is to be found in Berri Bible, and MS. 2897, British Museum. No. IV. is diapered chasing of a bowl from Nineveh, and exists nearly the same on carved work at Amiens Cathedral, and in XIV century MSS. No. III. from a horse-cloth in Ninevite sculpture, is very common in XIV century MSS. also. As a general rule, blue diapers are ruled in glossy black, purple diapers with glossy crimson lake.

II. Black square lining; dark squares blue smalt; stars in vermilion.

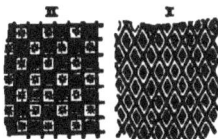

I. Diamond cross lining gold; quarries within, of cobalt blue.

IV. May be a burnished gold ground, and the pattern indented thereon, the cross-hatching being pricked in.

III. Blue ground; diamond black lining; white dotted quatrefoil; on purple ground, crimson lake lining, and cobalt dots.

VI. Horizontal and perpendicular lines squared in in black. Inner diamond ruled in pink, with pink circles. Middle diamond blue lines and circles, outer diamond green lines and circles. All dots white, all plain squares gold.

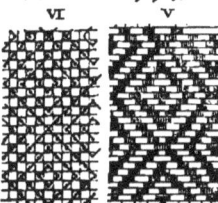

V. Blue ground; ruled in black parallel lines; the dark pattern in gold, or gold ground, and pattern in blue and green.

VIII. Purple ground, squared in with lake, white circles and dots, cobalt inner squares, and white dots, cobalt dots at intersections.

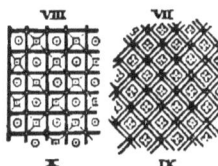

VII. Purple ground; cross bars in gold, with scarlet dots at intersections, quatrefoils in white, and red dot in centre.

X. Lined black; dark quarries —burnished gold; others smalt with white trident.

IX. Burnished gold ground and indented pattern.

XII. Purple ground, all gold lined, and gold corners and circles burnished.

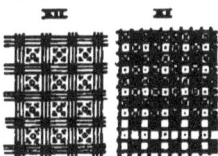

XI. Blue ground, dark squares gold, white circles and crossings, red dots.

Dark blue ground, gold or white scroll pattern, or gold ground and indented pattern

Dark purple or blue ground, with gold or white scroll pattern.

Indented pattern on gold ground.

* Some of these may be varied from blue to purple grounds, but must be finished then with lake lining, &c. I., II., III., VII., VIII. and IX., may be used in XIII century work, others for XIV. All may be enlarged for bold work.

Line Finishings.

HESE are generally finished as described for borders and initials, and in the same manner as some of the letters ' I ' which serve as borders, being in many respects the same, only laid horizontally. In some MSS. line finishings are put in in blue and red line scrolls with a little gold as p. 24 and Fig. 1, Plate III. Many examples of different styles on grounds are given at ends of chapters.

Having laid on all your colours, except the delicate white-line finishings, you should rub down in about a salt-spoonful of gum water in a small saucer, lamp black and Indian ink from the cakes, until the mixture is almost too thick to write with. You may freshen it by a little gum water or mix up new ink every three hours. With it, either by a pen, which suits you best, or by the brush, which you should early accustom yourself to, commence to outline your work with a firm flowing line of a requisite thickness, as in illustrations, with much the same intent as a school-boy tries to write a good bold firm large text, not shaky. Take elbow room on your table, and stand to it at a desk nearly as high as your breast; when tired you can sit at the work occasionally. Straight lines should be ruled with a bow pen, and small straight edge.

For the delicate white hair finishing, which is the last process, the smallest brush with short hair is best, a pen clogs up the colour.

Part V.

GILDING.

AVING already recapitulated the various preparations used in illumination, we will now treat of 'gilding,' which from its peculiar methods of working requires distinct explanation.

Gold Leaf and Shell Gold are the preparations of gold principally used in illumination.

Gold Leaf is the better kind to use, but its manipulation is extremely difficult. For the method of handling this material, during the process of gilding, we cannot do better than adopt the following rules, viz :

"Place the book of gold leaf on a level surface—quietly lift up the top paper, press it back and pass your finger along the 'bend,' if the gold be disturbed and uneven, blow it down smooth, and then with the right hand place it on the side of the leaf (close to the edge and but very little on the surface)—gently lift up the leaf, which will then 'float,' and so hold it that the current of air may waft it in such direction as will enable you to keep it spread well open—now let it slowly and steadily descend upon your work, whereon blow it down as flat and smooth as you can manage—do not be in a hurry to disengage the 'tip,' but take a piece of soft cotton-wool in your left hand and quietly 'pat' down the gold—then remove the tip.

The gilder's cushion is specially made for cutting gold upon. It is generally about eight inches by five in size, and is made by stretching a piece of soft wash-leather over a piece of board previously covered with baize.—The knife is generally termed a 'gilder's knife,' and is long and flexible.

There is an 'extra thick' gold leaf prepared expressly for illumination, and it is to be recommended as being much more manageable and easier to handle than the ordinary kind. Dentist's Gold can be handled and cut like thin paper, but it has the drawback of being very expensive.

There are two different preparations required for the proper application or working of Gold Leaf, viz : Water Mat Gold Size, and Raising Preparation. The first is used for causing gold leaf to adhere to a surface, and the second for raising an artificial surface on any portion of the work, afterwards to be covered with the leaf. These compositions or 'grounds' can be purchased ready prepared for use, and generally with printed instructions as to mode of application. The latter is often made ready to receive gold leaf, but is sometimes too absorbent, so that having been applied and afterwards wetted, it instantly drinks up the moisture, and refuses to take hold of the leaf. To remedy this, pass over a little white of egg made very thin with pure water.

For 'flat gilding,' with gold leaf, the following method may be adopted.

"Wet a brush (red sable in quill) and rub off plenty of your gold size, with which carefully trace a portion of your pattern, laying on a good body of the size. See that the design has been correctly drawn, and touch again with size, any part appearing weak or thin (keeping the surface level). Wait till the size be sticky, when breathe on it so that its surface may be uniformly damp (not wet), and then on it lay the gold leaf, which touch down lightly with a swan quill camel's hair brush, or a piece of soft wool. Wait till the whole be *thoroughly dry*, when lightly burnish the pattern, and then with a piece of wool carefully rub off the superfluous gold. Finally, burnish the pattern thoroughly."

In very particular work, when it is important that the gilding be unerringly level, it will be advisable, after having traced the pattern with a good body of size, to let it get thoroughly dry. Then level and burnish the surface of the size, and afterwards damp it (with water or diluted white of egg) to receive the leaf.

For ' raised gilding' it will be found advisable as a rule to use gold leaf (though shell-gold will often do very well), and the following directions may be followed.

" The ' raising ' or ' size' should be dropped or painted on in coats one after another, letting each coat dry before the next is applied. When a sufficient height—and very little will raise the gold—the gold may be cut. To cut the leaf properly, it should be laid flat, as previously directed, upon the cushion, and the knife-edge laid upon it in the direction the cut is to be made. This will steady the hand. The leaf is then easily and evenly cut, or rather sawn, by drawing the knife smartly but lightly down its edge. The knife-edge must be perfectly even, but not sharp, or it will tear the gold and cut the cushion. A few trials will give the requisite pressure and dexterity. When the pieces are all cut about the required size and shape—clean water, a sable pencil, and a gilding ' tip' should be at hand. Next wet the place for the gold, as far as the piece is to lie—generally a touch of the sable is sufficient, as the water spreads somewhat—and while quite wet, take up the piece of gold with the ' tip,' and lay it lightly over. No pressing is required. It will dry flat if properly done, and pressing will damage it, perhaps. All the raising may be gone over in this way by wetting the place and immediately laying on the gold. When dry or thereabouts, the burnisher is gently passed over with a motion like that used in shading with a lead pencil, only in long sweeps—the lighter the better at first. Then dust off the gold."

If this plan does not succeed, the fault will be chiefly in the sizing or raising preparation, as the operations are quite simple throughout.

Here again is another mode, suggestive, and well worth attention.

" After the raising preparation is laid on, and while it is drying, the gold should be cut into portions, ready for laying on. The plan is to cut off, with a pair of sharp scissors, the whole leaf or less if sufficient —paper and gold together—and then, with the gold still lying on the paper, cut through both to the shape and size wanted. This should be a little larger than the actual space to be covered. Then with a sable dipped in water, or diluted white of egg, as the nature of the ground may require, make the raising very wet if it be absorbent—just damp if not—and apply the gold by turning it over from the bit of paper as best may be managed. If skilfully done the gold clings at once, and needs no pressing, and looks best when dry, as pressure damages it and interferes with the burnishing. Lastly, after all has become quite dry— say next day to be sure—apply the burnisher, feeling the way gently until the polish comes."

Various receipts for the preparation of grounds and sizes, and for the method of gilding, are given in old tracts on illumination.

The following from a MSS. Harl. Coll. No. 3151, will be found very useful, as containing one of the secrets of smooth gilding ; and, as far as we know, never before published. We give the summary, the language being very rugged and prolix.

" First with a plumet (fine pen) trace the letter ; then the vignettes and images if you make any, and afterwards the portraits, with black ink. Make your size thus. First take chalk, and grind it on a muller till it is small and fine, and temper it with the glaire of an egg, water, and saffron. Mix with discretion so as not to make the size too wet or too stiff. Put the size in the ground or in a moist place for seven days. Stir it once a day. After straining, dry the size, and when dry ply it well. If it does not break it is good, if it break add more water. Next with a pencil of squirrel's tail, lay on a quantity rather substantially, and after that another, and let them dry. When dry, wet it slightly, and engross the letters, &c. with a sharp knife, and smooth the size. If required, set by a fire to dry. When dry, *burnish with a dog's tooth* set in a stick, and *when it is well burnished* and made even, it is ready to lay on the gold and silver. Silver size is made like that for gold except the saffron. Cut the gold with a sharp knife upon a pad made of calf's skin with flocks inside. Gently wet the size, and touch the pencil in the mouth to damp it for taking up the gold. Press down with a hare's tail, or pad of cotton, very slightly upon the size, and, when pressed, thoroughly dry it. If done before noon, it may lie till afternoon. Then burnish it ; if you burnish it while damp you will rub off all the gold. It may be dried and heated on the stove, then well burnished again. Do not burnish too long for fear of spoiling. When burnished, take a woollen or cotton cloth and rub away what is not held by the size—'and now prove thyself by experience, for use and exercise bringeth man to perfection.'"

To 'engross,' as mentioned above, means to rub or pare down the size, and smooth it ready for the burnisher: Ital. Sgrozzare. Size is really an 'assise,' that is, a setting or ground.

The part italicised is somewhat important, the cause of many failures in gilding being that the ground is rough, and the gold is licked off by the burnisher. Another common source of failure is, that the burnishing is attempted before the size is dry. This always ruins the gold.

We should remember then: 1. That the size must be pared and burnished *before* laying on the gold, and: 2. That we are not to burnish at all until the size is perfectly dry.

To Mr. T. G. Goodwin* we are indebted for the following interesting account of gilding as presumed to have been practised by the old illuminators.

"The gilding employed by the old illuminators was of two kinds, flat, and raised or embossed. The former was used for laying a large smooth surface for painting upon, and in many cases even for scrolls and ornaments, but chiefly for shading and putting in the highest lights on such colours as reds, browns, and yellows. The raised gilding was principally used for nimbi, or glories round the heads of saints, for lines, background letters, and small ornaments and leaves.

"This latter kind of gilding it is which is so much admired in all ancient MSS., and the art of which was for so long a time lost. Let us consider then how this was practised by the illuminators of the mediæval times. The first operation is to lay upon the letter or ornament to be so gilded a smooth plaster called Gesso, which is made thus: take equal parts of plaster of Paris and pipeclay reduced to powder; make a strong size of gelatine or fish-glue, mix this with the powder until it forms a stiff paste which may be worked with a brush. Add a little Armenian bole to colour and give richness. Proceed to lay a smooth surface of this Gesso upon your letter, adding gum-water to dilute it, if too stiff. You may raise the letter or ornament to any height you please, only be careful to keep the surface quite smooth and even. When this is dry and hard you may then proceed to size it, that is, to

* The author of the Appendix.

paint it over with a sticky substance to which the gold leaf will adhere. Now we will tell you what was the real size of the mediæval illuminators, and this shall be very serviceable to you. Take equal quantities of gum arabic and gum ammoniac reduced to powder, add half the quantity of Armenian bole, and mix them in a brass or glass mortar with very pure gum-water. This will form a thick paste, which you must moisten with your brush when you wish to use it. Upon the ground of Gesso you may now put a tolerably thick coating of this size, and then wait until it be ready to receive the gold leaf. This will generally be in about ten or fifteen minutes after sizing, but you may tell by the touch : if the size be dry and yet stick to the finger when you touch it, it is ready, but if it be not thoroughly dry and smears when touched, it will certainly spoil your gilding. When all is ready, breathe upon the letter, and then, having with a camel or sable hair pencil torn off a piece of gold leaf of the dimensions you wish, apply it softly to the size and it will adhere. Press it down firmly with a silken handkerchief and leave it for a while. Brush it softly with a large dry camel hair pencil to remove superfluous gold. If there be any defective parts, you must touch them with the size and regild as before. The last operation in gilding is to burnish the gold. This must be done about six or eight hours after gilding, if the weather be fine, but if it be damp or rainy you will have to wait longer. Place a plate of glass or some hard metal beneath the letter you have gilded, to produce a level surface. Be careful to see that your burnisher, which had best be an agate or dog's tooth set in a handle, be perfectly clean and free from dust, and to this end rub it up and down upon a piece of cloth before using. When you use it, move it slowly but with moderate pressure over the gold, turning it round in your hand as you proceed. If you wish to engrave patterns or lines upon the gold, you must use a proper agate graver with a blunt end for this purpose. You will also find a ball of cotton-wool covered with wash-leather very serviceable for, removing. superfluous pieces of gold. Now observe, with the Gesso, you may raise your gilding any height you please ; but if you are wise you will not make it of immoderate height, and this because not only will it give your work a tinselly and vain appearance, but also there is danger that the plaster chip. Let then your own good sense and ancient example be your guide.

"The other gilding we mentioned is called flat gilding, and consists of fine gold or rather bronze-powder mixed with sufficient gum-water to make it adhere, and laid on with a pencil. To do this only requires neatness and care ; even this kind of gilding may be raised if laid upon a ground of Gesso prepared as in embossed gold, but it is better suited for shading hatchings on garments, and scenery, and for very fine gold lines. You may relieve it occasionally by engraving spots and diapers with your graving-tool on the dead surface. This gold powder mixed with red or dark brown forms a very curious and rich pigment, and answers well for a deep ground colour. When mixed with light brown or sap-green it makes a good bronze, and may be heightened with pure gold or shaded in browns."

Shell Gold is very useful for small portions of 'flat' and occasionally of 'raised' gilding. It is a very good substitute for leaf, and is easy of application.—In some old treatises it is the only sort allowed. The following directions may be adopted.

"The portions intended to be gilt should be painted over with proper size or with white of egg well frothed and diluted with a little water. When nearly dry it may again be painted over with shell gold. This, when dry, is burnished with the gentlest possible touch of the burnisher."

For large surfaces of flat gilding, gold paper will be found very useful. It can be obtained ready prepared with a strong solution of glue and sugar on the back, which, when damped, causes the paper to adhere firmly to the surface of the work. When laid down, a piece of smooth paper should be placed on it, and a napkin or the edge of a flat rule firmly passed over.

When a 'white metal' is required, we would recommend shell platina as being less liable to tarnish than any other white preparation. These 'shell' compositions have a little water applied to them, and being stirred up with the brush, are then ready for use.

The metallic paints are generally prepared in two bottles, one of powder and the other of liquid, and require to be mixed. The metallic powders and inks are sold ready mixed, and have only to be well shaken up before being used. In all these compositions the rule applies of suffering them to get perfectly dry before burnishing.

KEEPING in mind it is hoped, all that has been previously stated, especially in attempts of distinguishing the best styles, and basing your practice on choice examples recommended to you, we may, in concluding, merely add one or two general remarks.

I. That great proficience can only be attained by copying from good examples, repeatedly and continually, until we have sufficient stock to work upon. It is absurd to think that we can begin to invent or apply ourselves to original composition, before we are possessed of the materials, in the shape of an extensive knowledge of what has been already done, and especially before we have gained sufficient facility of practice. From what we have already noticed we observe:

II. That a careful outline is an imperative necessity, and that the beauty of this outline depends on the flow of its curves, which should be subtle and changeful in their character, and in general, proceeding from the main line ; and that all component parts of a composition, combining border, miniature, or capital, must have some connecting features in detail. That too much delicacy cannot be bestowed upon the graceful outlining of figures, in expression of force and management of draperies ; and also in forms of leaves, buds, and even grotesque animals.

III. That in true illumination the principal features depend on contrast and variety, and are light or relieved on dark grounds, and are gradated by darker or lighter shades of colour. That although science is a great help in the arrangement of colour, nevertheless, *whenever* colours look well together, they are right, whether scientifically so or not, according to the present state of colour science.

IV. That in order to give greater effect to colours, and make them tell, a fine boundary of black and white should be used. If no ground colour is used a black outline is best, not only for the sake of colour, but to give force to the design. On gold, no outline is absolutely required, and a gold outline will harmonize with any ground whatever.

V. That if rich deep compound colour is wanted, it is best produced by mixing opaque with transparent colour, not two transparents which give weakness, nor two opaques which give heaviness of tint.

That to give the bloom to flat grounds, the colours must be paled with white, though never in such quantity as to overpower the after decoration.

That to give solidity to these grounds, and to preserve fugitive colours from destruction, it was customary with the old illuminators to mix their colours up with glair of egg, more or less thinned with water. This is better than either gum or oil for paper and vellum.

VI. That it is essential to get into cleanly and orderly habits of working. Some artists despise methods, and think it is " artistic" to be slovenly, but you will produce more work, and better work, and save much time and annoyance by good arrangements.

VII. In conclusion, let us remember that for the higher

portions of our art, our best teacher will be Nature, in
adapting her fine ever-varying suggestive forms, whether
in the human figure, or in graceful drapery; in animals
and birds; in flowers and trees with the lovely colours
and beautiful curves from branches and leaves. Sometimes suggestions from such objects strike us most forcibly when least sought for. We may accidentally see a
tender expression of face, a graceful fold of drapery, or
a fine curve on a hound's back, or on neck of a swan or
dove, or in a branch of ivy or a leaf, or leaning over of
a flower, which we should retain in our memory, or note
down; for imagination may not be so strong as to supply
these impressions unless studied and sketched from
nature.

But in these things, let us not forget that experience
comes by practice and copying of many examples, to
which must be added patience, careful observation, and
" taking notes."

Appendix.

BY T. G. GOODWIN, B.A.

*LANDSCAPE, FIGURE, STYLES,
COPYING AND DESIGN.*

OUNG or inexperienc-
ed learners, earnest in
perfecting themselves in
the Art of Illumination,
may reasonably be sup-
posed to seek informa-
tion upon the character
of Mediæval Landscape and Figure, as applied to design—which we will endeavour
to explain.

The unsatisfactory character of most
modern illuminations mainly appears to
result from the neglect of certain broad
rules of composition,* belonging equally
to all Arts, and never transgressed without lamentable failure and disappointment.

---

* It may be justly added also, from students trust-
ing to modern weak published designs, rather than
studying the excellent ancient examples for themselves.

It will, therefore, be of importance to impress upon the pupil a few simple rules, which may to some extent serve to keep him from like mistakes.

Landscape.

TO give you elementary instruction and exercises in this branch of the art, is not my intention, nor will my limits allow me, even if my inclination led me, to discuss the principles on which landscapes and figures must be drawn. But supposing you to have already acquired some degree of proficiency in this art, and to possess a correct knowledge of drawing and colouring, from the many excellent works that already exist on this subject.* I shall proceed to lay before you what I believe to be the peculiar characteristics of the paintings we find adorning the pages of manuscripts, more especially such as relate to their principles of shadings, and the tints most frequently used. It will at once be seen how very great is the difference between a landscape painted with transparent and one painted with opaque colour, that is, colour mixed with Chinese white. By the latter process, colours are obtained firmer, purer, and brighter in tint than it is possible to effect with transparent colours. It is no slight gain also to be able to strike one colour, even if lighter, over another, as by this means very subtle and delicate effects are produced. It must, however, always be borne in mind that the upper tint must itself be pure, that is, unmixed.

Let us begin then in an orderly manner to consider the points in an illuminated landscape, and first, the sky. There is a wonderful sameness perceivable in all the mediæval skies. They are generally a broad, open expanse of blue, delicately graduated, fading into white

* Vide J. D. Harding's "Elementary Art," Winsor and Newton.

down to the horizon, and broken with a few occasional wreaths of cloud, the lights of which are put in white, and the shaded parts either in darker ultramarine or indigo.* The rays of the sun, stars, or meteors, are always represented by actual flat gilding, which has a very beautiful appearance. Skies when masterly and gently graduated, are almost always the best parts of the landscape, and have an inexpressibly calm and soothing appearance.

Mountains and rocks are occasionally given in somewhat arbitrary colours, but if in the extreme distance, were picked out in dark blue and white. The lights, especially on rugged cliffs and such places, were inserted in gold. Houses, towns, and buildings in the distance, we find generally softly pencilled in with a dark outline, so as to stand in bold relief against the pale blue sky. Lilac too is a favourite colour for distant features. For trees, the most common colours were a dark emerald, sap, and cobalt green, indigo being much used for shades, while for grass, emerald and green bice are most often found, and the tufts of grass expressed in white or yellow. All these, however, are by no means intended for rules, as they can only be learnt by long and careful examination of MSS. of different dates.†

Figure.

AINTING of figures requires a course somewhat similar to the foregoing. The faces, hands, and flesh generally, after being first sketched and outlined with a pen and ink, were either elaborately finished in solid colours, with much shading and stippling, or slightly tinted

* In XII, XIII, XIV and XV century MSS., clouds were painted in a rounded zig-zag white shaded form. This may be seen (but faintly in such small space) in letter A, Plate II. and illustration at p. 71. The diapered backgrounds of letters and miniatures are conventionalized treatments of unbroken skies, or hangings.

† See "Modern Painters," Vols. III. and IV., for some useful instruction and illustrations on the mediæval treatment of skies, rocks, trees, &c.

and the features put in brown; the pure surface of the vellum was left for the flesh, but shaded with a brown heightened with white. But in all MSS. of the best date, that is from about 1250 to the end of the fourteenth century, you will find the utmost delicacy in delineation, and tenderness in colouring faces.* There is every certainty that in most cases they were either copied from portraits already in existence, or from the personages themselves, whom the illuminator wished to represent.

Very elaborate and beautiful also were the costumes, and the general disposition and flow of drapery cannot be too much studied or followed. The many heraldic bearings, so common upon the dresses of that period, were all delineated with the utmost accuracy and finish; the artists of those days evidently took great pleasure in elaborate patterns and diapers on garments, and displayed the highest amount of patience in the labour bestowed thereupon.† In a word, and it is a fact which you cannot dwell upon too much, there is nothing like slovenliness or niggardliness to be found in the work of the best periods. Labour was given cheerfully and ungrudgingly, and the artist's heart and soul were in his work. How different is this from those later times, when first the type and then the border itself were printed by machinery, and then left to be coloured by hand!

Doubtless, much of the brilliancy of a mediæval miniature is owing to the bright gold hatchings and shading which adorn trees, architecture, and costume. To prove this, try to paint a miniature without the aid of these and see what the effect would be. But observe, do not run into the other extreme, and gild your pictures all

---

* By close examination of many MSS. of that period, the whole head is painted in with white on the vellum, and only the cheeks touched with rose colour. The eyes, nose and mouth are delicately and masterly drawn in, and the hair so skilfully as almost to be inimitable.

† Admirably and most variedly delineated in the Berri Bible, and 2897, British Museum.

over, for your gold should be used with the greatest
care, and never but with a special end. Try to make
its force felt and understood, for this will show its true
value.

We must bear in mind that the mediæval illuminator
painted in subjection to a certain set of conventional ideas
which was handed down from age to age. This will
account in a measure for the great similarity in works of
this kind. Thus, there was a certain type of an angel,
another of a prophet, another of a martyr while the
numerous emblems and accessories of that nature, which
were introduced, naturally tied down the painter to one
style.

Now I do not mean to say that their landscape or
figure drawing is in all respects to be literally followed
in these days, but I do most decidedly assert that it
is by no means to be altogether rejected or despised as
something barbarous, but that upon it we must found
our practice, if at all we wish to illuminate. Rude and
ill-drawn in many respects it often was, the perspective
was frequently absurd, and nothing is easier than for
persons *who could not imitate its beauties to ridicule its
defects.* And it is these beauties we must study—these
defects we must avoid. I would not have you in your
practice of illumination, lay aside one single true principle
of drawing, or relinquish one single improvement in
mechanical helps which has been discovered; I would not
have you paint with anything less than your very best in
knowledge, time, and care, nor would I have you voluntarily paint in a manner which you inwardly feel to be
grotesque and absurd. No; give your best work; but do
not despise the labours of others, and try if you can to
imitate their many beauties. I will maintain that there
is more true celestial beauty in an angel clothed and
represented as Angelico has his angels, than in a naked
sprawling cupid kicking in the midst of flowers and
fruits, which in the days of the Renaissance passed for an

inhabitant of heaven ; and that there was at least as
much propriety in painting the Blessed Virgin, that Holy
Maid, with her glowing nimbus, and the colours which
have of old been appropriated to her, as in portraying
her as a modern young lady decked in white satin,
and with the gaudy decorations of this world's invention.

For purity of sentiment, simple severity of treatment,
and tenderness of execution, the MSS. of the XIII and
XIV century are unrivalled, and it is to these that we
must look for models and guides.   Make it your rule then
to follow them in all points which do not actually offend
against the rules of nature, and which the requirements
of Art demand, and, believe me, you cannot go far
wrong.

Before I leave the subject of figure drawing, I would
strongly recommend all who intend applying it to illumination, to study and endeavour to become acquainted
with the proper emblems of the saints, and such other
minutiæ as show an accurate and painstaking mind.
Believe me, you will not regret this, nor, indeed any
other worthily bestowed labour.

Styles.

LET us now advance, first to copying a page
of some MS., and then, as I trust you will
some day do, to design one in the style we
may fix upon as the most beautiful and
truthful.   But first, I must detain you a few minutes
with a brief notice of the various styles, and the order
in which they succeed each other.

No one who has carefully considered the subject can
have failed to remark how exactly the advance and
development of illumination kept pace with that of

architecture. Nor is this surprising, when we reflect that
in the Middle Ages the practice of the various branches
of Art was generally united in one person, and that the
same man was frequently illuminator, artist, sculptor,
and architect.* It is not surprising then that we find
the same characteristics in ornamental detail and general
effect exhibited in the MSS. and architecture of the early
and Middle Ages, or that one idea is carried out in each.
I shall therefore divide the styles of illumination into
periods contemporaneous with the various developments
of architecture, and if you will take the trouble to examine carefully each style as we proceed, and mutually
to compare them, I think that you will recognize the
truth of my assertion.

The first style I shall notice, is that which prevailed
more or less from the viii to the middle of xii centuries. I say "more or less," because the gradations
from the Byzantine through the Anglo-Saxon and Anglo-
Norman Schools to the Transitional style, which immediately preceded the early English, are very decided,
and in some instances require very minute examination
to trace any connecting link between them. Still, just
as in the architecture of this date there is to a great
extent the same characteristics, so is it with the illuminations. The rude but intricate and involved foliage,
frequently mingled with grotesque and monstrous heads
of figures, which form so great a part of the architectural
decoration of this period, is accurately reproduced in the
illuminations, indeed, many of the ornaments of capitals
and mouldings seem taken from the pages of a MS.
Again, the leading features of the circular and segmental
arches are largely employed in MS. decoration, and all
the characteristics of grotesqueness and somewhat heavy

---

* If readers have seen the "Sketch-book of Villars d'Honnecourt,"
an architect of the xiii century, edited by Professor Willis, they will
see that his style of drawing is the same as in illumination and glass-
painting.

grandeur, are to be found more prominently here than anywhere else, and at once prove the connecting link between two Arts.

The style called Semi-Norman or Transitional appears to have flourished during the latter half of the xii century, at which time a decided change is perceptible in the illuminations. The solidity of the former style is gradually exchanged for a lighter and more elegant taste, corresponding very nearly to the change effected at that time in architecture by the use of intersecting semi-circular arches, and of the lancet form. The grotesque animals in initial letters, are by degrees exchanged for a broad entwisted foliage, in which, however, they were occasionally allowed to break through. The colouring of this period is somewhat lighter and more brilliant. Altogether there is a decided advance to the delicacy and beauty of the next or first pointed period.

Again, during the xiii century, illumination and architecture advance hand in hand. It is impossible for any one who has at all studied either the MSS., or the buildings of this period, not to remark the great similarity in general effect. In architecture we find bold and deep mouldings, severely pointed windows and arches, and simple but stiff foliage, in which the stalk is made a very prominent object. Just so in illumination, the broad foliage of former styles is exchanged for interlacing stalks and foliage, a combination which, though most elegant in itself, has all the stiffness and want of elasticity which is to be found in sculpture. Exact parallels too may be found in the pages of xiii century MSS. to dog-tooth moulding, the trefoil shape and crockets —all of which seem to have made their first appearance at this time.

Again, as we find the purest and most beautiful architecture existing from the latter part of the xiii century, when the first was gradually merging into the second pointed style, to the middle of the xiv, when it began to lose its

power, and decline into the corrupt luxuriance of the xv or third pointed—just in like manner the most exquisite specimens of illumination and the greatest illuminators, including Giotto, Angelico, and Frederigo d'Agobbio, are to be found. Purity of conception, grace in design, beauty of colour, and inimitable delicacy in execution, stamp the Art of this period, and render it in all its branches the most valuable guide for students. You cannot dwell too much on, or study the MSS. of this date too eagerly. Embrace every opportunity of seeing or copying them, and this trustfully and enthusiastically. I shall reserve what more I have to say on this period until I come to speak of Design.

The gradual corruption and decline of the xv century architecture is well known, as well as its subsequent overthrow by the so-called Revival of the Classical Styles. A change so great as this, which infected nearly all Europe, could not, of course, overlook the MSS. and illuminations, and we soon find the pestilential influence extending itself through their pages. It is, however, by degrees that all evil works, and so in many MSS. it is scarcely perceptible. How-

ever, in time the end came, and illumination over-bur-
dened with so-called classical conceits and inventions,
which it had learnt and copied from the degraded Art of
the day, was in the end itself deserted by its treacherous
ally, and finally flung away as one of the relics of the
barbarous ages, which it was the boast of the Renaissance
that it came to destroy.

Copying.

SUPPOSING that about to be copied, to
belong to a MS. of the XIV century or
thereabouts. First having cut a piece of
vellum, (if you intend to use vellum,) the
size the page is to be, prepare it by rubbing
it with pounce. Then, having stretched it
out upon a board, rule very lightly the
lines for the margin, type, and initial letter. If the
border be open, that is, upon a white ground, as most
of this period were, the outer marginal line will have
to be erased, so that it had better only be ruled in pencil,
but your other lines may be ruled with light Indian ink,
using, of course, a proper drawing pen.

You will next proceed to print in the type proper to
the date, sketching the capitals, which you will finish
afterwards. The next thing to be done is to copy the
large initial letter, which you must do with great care,
testing its accuracy by a tracing, and finally rendering
its outline permanent by Indian ink and a pen. If
there is to be any picture, you will sketch it in with
pen and ink. You cannot be too delicate in your outline.

Last of all you will copy the border, and that in the
following way. Fix your eye upon some prominent
portion of the border in your copy, and having ascertained
its exact position and dimensions, proceed to mark it out

upon the surface of your vellum. Measure again the distance from this to the next most prominent feature, and so on in like manner until you have all the most important parts fixed in their proper places.

Now advance to the subordinate ornamental detail and gradually fill that in, dividing your work into small portions and taking the greatest care to have all correct. Do not rest till you have a literal facsimile of the original, and then fix it with diluted Indian ink outlines.

You may now proceed to colour. First of all observe that if the vellum be at all greasy, you had better mix a little liquid ox-gall and water with your colours, which will enable you to paint with ease and certainty. You must commence by applying all the gilding throughout the page first, as otherwise you would rub your colours when you began to burnish. You may then finish the initial letter, using your colours as already shown you. Next begin to colour the border, applying one tint wherever it is wanted all through it, then finishing the next, and so on until you have all the colours laid on their proper places. These you will proceed to shade and ornament in solid Chinese white or gold; any little figures also or grotesques you may now complete, including of course the terminal line, generally of gold and colour, which encloses the type. When you have all these quite finished and really accurate, both in shape and colour, if there be a picture you may advance to that, but if not, you will put in any dots of colour or flat gold which may happen to adorn the background.

This will conclude the operation of copying, and you may finally erase delicately with bread any spots or marks which have occurred in the course of your drawing.

If you are diligent, you will make many such copies as these to keep by you for reference, taking care to ascertain and attach to them their proper date, which will much enhance their value. You will also find fac-

similes of small details copied into a note-book of great
service.

Design.

ESIGN of a proper kind, the reader
need not be told, is of all the aims of
art, that which is the most difficult, and
in which skill is always desirable.

The first thing which appears necessary to do, is to fix upon some already
existing style of illumination to serve as a basis
for any intended developments. Of course this style
must naturally be the best and purest, and that is,
as I have often said before, and as I firmly believe, that which prevailed from the middle of the XIII
to the end of the XIV century. But, perhaps, you will
ask why I say so much in favour of illuminations of this
period. and it is but fair that I should give you my reasons for so doing, which are these.

The illuminations of this time appear to me to fulfil
perfectly the purposes for which they were designed, and
the conditions under which they ought to be executed.
The ornament is strictly connected with the construction,
which you will see is one of the most important principles. There is nothing forced or overburdened in the
introduction of ornamental detail, and in itself it is of
the purest and simplest type and guided by a healthy
conventionalism of nature. You will find no obtrusive-
ness of individual detail, but all tends to form one
harmonious whole, which is wonderfully suited for the
purpose for which it was intended.

Assuming then the acceptation of this style, I will now
proceed to consider what I think to be the great principles
on which design as applied to illuminations must be con-
ducted. The first is, that *ornament should spring from the
necessity of construction*, in other words that there should be

a good and sufficient reason for introducing it. Now the
"necessity" of a page is without doubt the terminal line,
that is, the line which confines the type. Our ornament
therefore must either spring from or in some way identify
itself with this line. Now in the early times the terminal
line was itself the border and was consequently broad and
much ornamented, so we find Saxon MSS. written under
semi-circular arches. Next the initial letter was carried
all down the page, another application of the same idea,
and lastly both initial and all other ornamental detail
immediately sprung from this line and connected themselves with it. But whenever the border is entirely
detached from this line it is wrong both in itself and its
tendencies. Make it your invariable rule then not to
detach your ornament from your construction, but to
render the latter as beautiful as you possibly can.

But you will ask of what nature ornament is to be, and
this leads us to the second rule, that *ornamentation must consist of conventionalized representation of natural objects.** All
true beauty consists in the representation of, or is derived

---

* Students must be puzzled for a reason for the sharp prickly edgings
round curved stems, especially in XIII century work, and in the "ivy"
borders, as in Plates II., III., IV., and indeed in almost all. In a previous note, it was stated, that the XIII century illuminator was content
to give us suggestive design from nature's beautiful forms. I know not
where the ancient artist got this prickly design, but if you pick up an
old hard dried thorn leaf, like Fig. 3, Plate II. and note the fine central
curve-vein in it, and can see that it is beautiful, you may ask yourself,
how shall I apply this in some design? It might do as a stem, like an ivy
stem, and a bud at end; but then, I want all these fine out and in
curves with the sharp points—these shall form a kind of background.
Study it closer, or turn it round and you will see the leaf's edges have
crumpled over and left a white edging which is not unlike the delicate
hair line within the black. To carry it farther, make white curved lines
in harmony with the great main vein to answer for the small thread
veins of the leaf, and you conventionalize it in *form* to Fig. 4, Plate II.
Then to conventionalize the *colour* you make the stem pink on a blue
or gold ground, or pale blue on purple or gold. Keeping the fine first
suggestive curve in view as a stem or branch, why not fill it up by a bit
of fun, and paint the portrait of the favourite turnspit dog on it, or a
crow, or a rabbit, or a monkey, or some fantastic imaginary fellow.

from, natural objects. To this rule there is no exception.
But nature may be represented in two ways, either by
realizing her as far as our means will possibly allow, or
by conventionalism. In realism, we endeavour to obtain
a literal copy of the object to be represented and set no
bounds to our pursuit of this. In conventionalism, we
beforehand arrange certain limits at which to stop,
and then get as much of nature as we can within those
limits.

Now the reasons why the former system, as applied to
illumination, is wrong appear to be these. *First*.—
Flowers and leaves may be portrayed in a page in three
different ways—as a wreath disconnected from the marginal line, or as springing from it, or lastly in the shape
of single flowers, or groups of flowers, spotted over the
page. The first and third of these ways are ex-
cluded by the rule which directs all ornament to proceed
from the terminal line, even if they were admissable
otherwise, which they are not. The second condition of
the wreath then is left. But how is this wreath, upon
which you have expended the greatest care to bring it to
an exact resemblance of nature, to be supported, in fact
what does it do there? If we are to be perfectly natural
we must assign it some support, as wreaths are not wont
to hang self supported in vacuity. Unless you are prepared to encounter all the elaboracies of a back-ground,
and that too of a probable kind, the unhappy wreath both
figuratively and literally falls to the ground, for of course
the marginal line itself cannot afford any support consistently with the natural principles we are supposed to
advocate. *Secondly.*—Absolute realism of foliage and
flowers would after all reduce illumination to simple
flower painting. By using the word "reduce," I do not
mean anything against the art of flower painting, on the
contrary, it is most beautiful *in its proper place*. Again,
if we may not stop short in our absolute representation of
nature, we cannot introduce arbitrary forms such as

scrolls and grotesque ornaments consistently with our principles. *Thirdly.*—A wreath or arrangement of natural flowers should be wanting in that repose which we must endeavour to attain. There would be a struggle among the details for prominency which would end in utterly destroying all feeling of repose, and, distracting the attention entirely from the type, would give an undue preeminency to the border. For all these reasons then we must conclude against naturalism.

Now in conventionalism it is to a great extent optional how far you will realize your flowers or leaves. You may do it more or less as you feel the occasion requires. The principle appears to be to seize upon the leading characteristics of the flower you wish to represent, and then to add as much of the rest as you can consistently with your subject. Thus, you may either represent a rose as an arrangement of five leaves of a certain shape and colour round a yellow or gold central spot—as was the general mediæval type—or you may go somewhat nearer the reality and add a few more petals, &c., so as to bring it to a closer resemblance. You must arrange this with yourself, but as a rule observe that the more you realize any flower the more you must proportionably increase the quantity of conventional ornament around it, so as to make it evident that you had voluntarily set yourself limits which you did not choose to pass. If you try to regard the flowers you wish to introduce as so many brilliant bits of colour to be arranged as best may suit your design, this idea will preserve you from any mawkish sentimentality, and will keep up in you a healthy and vigorous conventionalism.

While I am on this subject, I will take the opportunity to advise you to make great use of leaves in your designs. Wonderful and perfect as all nature's work is, yet it seems as if the stamp of perfection and divine beauty were more strongly impressed on leaves than on any other of her productions. The

thousand changing forms of beauty with which she clothes
the woods, the banks, and the very ground we tread on,
ought to be to all, but especially to lovers of beauty and
truth, objects of the purest joy and delight. Make very
frequent use of them in designing, for they ever
have been sources of the best and most heavenly
beauty. Remember always that in painting them it is
*as much important to have the form and outline quite
right and true* as to imitate or approach their colour,
which may be left arbitrary. Remember also that their
power will be better felt by a somewhat sparing use of
them, I mean as regards not over-crowding your page, so
that though you may employ many leaves, have but few
of each.

Play your flowers and leaves about boldly and freely
and put them in, like bright gems of colour, just as
you need them. Let your gold scroll and branch work
spring out boldly from the terminal line and at once show
its connexion with it.* If you want to introduce any animals or birds, as for the sake of grotesqueness you
ought to do, do not trouble yourself too much about their
species, or try to count all the spots on a leopard's coat
before you paint it. Let them tell their own tale simply
and boldly, and it will not matter that they are not anato-
mically exact zoological specimens. Do not however go to
the other extreme and violate nature's laws altogether, as
this, I firmly believe, is allowable only in heraldry.

The third and last rule of design is, *that there should be a
general purpose and meaning running throughout the ornamental detail.* You should endeavour to carry out some
idea in each border, and to this end should reflect well,
first what idea you wish to convey, next how and by what
means you may best convey it. I will not say that your
meaning will at once be plain to every one, nor indeed is
it likely to be so to more than a few, but still the working

* As in borders of Plates IV. and VI. Letter H, Plate V., and
other illustrations.

with a deliberate idea in your mind will give a unity and a completeness to your design, which will entirely be wanting to one worked out at random, or with a view only to prettiness—the most noxious idea it is possible to conceive, and the rock upon which nearly all modern illuminators make shipwreck. Accustom yourself to ask not " is this pretty?" but, is it "*right?*" and this habit will, I think be a safeguard to you.

We have now considered what, I believe, are the principles of design as applied to this Art. There are it is true many other points on which I could speak, but as my space will not allow it, I have chosen those which are the most important, and against which there is to beginners the greatest temptation to err. Careful study of the best MSS. you have any opportunity of seeing, must be your guide on other points. But as an encouragement let me tell you, that if you have any talent for design, and will take the trouble diligently to *think* over the rules I have given, and try them by such xiv century examples as may fall in your way, I do not think that you can go wrong in any material point. Differences of opinion and taste there must always be, but as long as we grasp the truth and resolutely cling to our landmarks, our steps cannot go far astray.

Practical Notes.

J. J. LAING.

DOUBTLESS readers will think that this Manual is a true example that "of making books there is no end." But with a view towards rendering it as efficient as possible, and of furnishing instruction in the best methods of practical working in the Art of Illumination, the reviser, as expressly desired, adds a few concise notes from his own system of working, as applicable to XIII and XIV century styles—information which perhaps may prove useful in addition to the instruction so ably given by Messrs. Bradley and Goodwin.

G

Paper or Vellum.

F the piece of work to be done is to be on paper or vellum, larger than imperial 8vo., and is to be coloured, it should be damped and stretched on a small drawing board.   Damp freely the worst side of the vellum, or that side of the paper through which you read the maker's name reversed, with a broad flat brush and water, (do not rub it with a sponge) ; wait till the damp gloss disappears, and then turn the paper on the board, (the damp side being under) and quickly apply a little fine warm glue to the edge of the paper, about quarter of an inch broad, first on two long sides and then on top and bottom, and rub down with a paper cutter, taking care that the paper sticks to the board.   Then give all the surface a slight wash with water tinctured with a few drops of ox-gall, and leave it to dry, but away from the fire, and in about two hours it will be ready for use.   Thick Bristol boards may not require stretching—thin may.   In every case, this damping at least is always better for colour. Vellum and solid books of drawing paper can be had ready stretched.

Sketching and Tracing.

F course, it is supposed that you can draw a little before you attempt illumination.   This is the more requisite, as in our Public libraries no tracing from coloured MSS. is allowed, but they must be copied through glass cases—so the sooner you begin to draw carefully the better. Under Part III, "How to set to work," are given preliminary instructions we need not repeat here. One leading rule in copying is to judge of the quantity or form of one piece of colour contained within the lines. If you are allowed to measure on a slip of paper the height and width of a piece you wish to copy, then take the distance of ruled lines, or count how many ruled lines are within margin, and thus you have a scale to go by, as in Plate VII., where the eighth line gives lower line of 'B;' sixth line, top of chorister's head; a little more than sixth line from bottom, gives top of monkey's head; and end of right margin line gives the bottom knot of branch; and then fill in the intervening parts after fixing these landmarks as it were. By using these ruling lines as a scale, larger or smaller, you might make a true proportioned drawing of any size. On vellum and fine paper, you should sketch with an HH pencil very lightly—do not erase any mistake if colour is to go above it—but if not, efface it gently with a knife or eraser, not with water, or Indian rubber, or bread, which the two latter may do, however, on paper or Bristol board—water leaves a dull blot unless the whole surface has been damped.

Tracing, when you have opportunity, is a simple process, and should be carefully done with a sharp H pencil. Small parts such as heads, &c., should not have all the minutiæ traced, but should be sketched apart larger on your tracing.

Transferring.

IX your tracing paper by a pin-head size drop of gum water to your vellum or paper at two or three corners near or beyond margin, so as to leave no blur. Then lay your transfer paper (black I find best, and less greasy than red) beneath it, and trace over your drawing, not too hard, with ivory tracer, lifting it up to see if you are doing it right. If the tracing is indistinct over the black paper, slip in between them a leaf of *thin* Indian post paper. Before removing the tracing, see that all is traced in correctly.

Colouring in Masses or Grounds.

ENERALLY in xii, xiii, and early xiv century work, you will find few colours used. Smalt, and red-purple, are used in large quantities say in backgrounds and dresses, (gold only rarely) ; scarlet, green, white, cobalt and pale pink, or pale rose colour, very seldom ; yellow in small quantities. So it was in good glass painting of this date. In later styles, yellow, indifferent greens, and reds, were used in masses, and gold for MSS. was used lavishly, and so became gaudy.

To match a colour we should paint a little on a slip of paper, and compare it with a part of the copy, shutting out from consideration all other colours and effects, so as to make our blue like the original, without its neighbouring purple, and black outline and white veining, all of which alter its first effect.

In mixing up the colours, it would be advisable to adopt the plan of never dipping a brush into the gallipots of colour, if these are used, but cutting a small quantity out with a knife, and mixing it with water on your slab or palette, always cleansing your knife thereafter—thus keeping the colours pure, and not wasting the points of brushes.

Mix up two or three of the principal colours at once, such as the blue and red-purple, and scarlet, as then they have more time to dissolve thoroughly ; and every time you take up any colour, stir and mix it well.

Be provided with at least four brushes, and two glasses of water, one to wash brushes in, the other to supply clean water.

BLUE FOR GROUNDS.—All parts to be blue, are the better for an additional slight wash of water, as blue is granulous and most difficult to manage. It should always be mixed with a little Chinese white, and very little gum—very little, so as to give *no* gloss. Smalt or ultramarine, or both mixed, is the nearest approach to the XIII and XIV century blue for grounds. Finish in all blue parts of the ground throughout the whole drawing.

PURPLE FOR GROUNDS.—It has been previously stated in a note at p. 12, Part I., that of all colours, purple is the most puzzling to match, from its varied tones. One good purple you may compound with purple-lake, a little orange and Chinese white; a touch of carmine or scarlet lake may be required. Sometimes the ground is of a uniform tone, other times as if a glossy scarlet lake or orange had been dashed here and there on the wet colour—thus making it cloudy, as in ' Vegetius' MS. Occasionally pink madder or brown madder may be needed to make up the right colour.

Blue and purple diapers have their ground colours of the same tone as mixed up for other grounds that have no diapers, although they may appear paler; which is by reason of the effect of the netted work.

When you see a fine rich purple, you should endeavour to fix it in your memory noting it as like the ' Vegetius' cloudy purple; or MS. 17,341 carmine-purple; or the pink-purple of St. Dunstan MS., because you may not have your colours by you to mix up a certain purple, and a description may have to suffice.

In working with these or any large masses of colour, lay your board flat and do not make long strokes with the brush, because the lower part of the stroke may get dry before your wet tint comes down to it from the top, but work the colour in quickly, keeping it all moist or floating.

GOLD.—Some illuminators advise that the gold should be laid on before all colouring, so as to prevent glazing of other colours by burnishing; but if you look at ancient MSS., you will see that the gold was almost the last process, because the finished black lining in some

parts is accidentally smeared over with gold or gold leaf,
which was done by the special " gilder." By burnishing
your paper near the gold before other colours are laid
on, the surface will be smoothed, but the glazing is
immaterial, and soon disappears. Avoid, however,
glazing the parts of vellum or paper which will not
have any colour.

As full instructions have already been given in preceding part of this work on the different methods of
using gold, it would be almost useless repetition to add
anything here. We may mention, however, that the
application of gold requires long and patient practice.
For general illuminations, shell gold is recommended
in preference to the troublesome work of gold leaf.
All gold, whether flat or raised, is the better for a composition underneath to hold the gold. A good method is
to paint on first some water gold mat size, over all portions to be gilded, then while this is sticky, or damp from
your breathing on it, to lay on the raising ground, (keeping the board flat) in one uniform layer; and after
allowing it to dry for an hour, to gently scrape off any
grits with your eraser, and then apply the shell gold.
It is not advisable to raise gold too much, as is the rage.
Many MSS. appear to have raised gold when it is only
the vellum with a slight under composition, puffed up by
the friction of burnishing. The stems, or leaves and
figures with gold grounds, by right should be more relieved than the gold. But nimbi, crowns, balls of gold,
and seats, narrow border bars, &c., should be raised.
The shell gold should be well mixed up with a brush with
but a few drops of water:—the largest shell may contain
five or six, a small shell three drops. In applying it,
do not wash the brush up and down overmuch, else you
stir up the raising composition, and then it becomes
with the gold only gold mud. But gently tint it on
with a soft camel hair pencil. If it gets bright in a
minute, it will succeed. Do not put on one wash

or coat of gold over another—if not right at the first, it will not mend easily. You must practice more: it would take long to tell what this practice consists of, but it includes noting if your composition ground has been sufficiently dry, and if it was a fair or a wet day, or the room cold or warm, and if you had mixed your shell gold sufficiently, and not made it too watery, or laid it on too freely.

Burnishing.

HOOSE a sharp or a broad burnisher as convenient. The success of burnishing results from experience or practice, and is influenced by the heat of the room drying the gold quickly or slowly: but generally by the time you may have gone over all the gold parts down to the bottom of a page, the first gilded part will be dry for burnishing—this you may try gently with the burnisher, and if it glides about easily, it is ready. The gold should not be too dry, for if slightly damp it adheres to the composition better. Indenting a pattern on gold is best done by the point of an ivory tracer, or by a thick darning needle. The indented lines may be drawn with the worn point of a pen-knife, not too sharp to cut. Some gold grounds, as in background of 'B,' Plate VII., are first all burnished, and then a pattern as the quatrefoil is painted over with shell gold, and left dead or unburnished.

Note.—All gold should be burnished before any green, (especially emerald or oxide of chromium) is painted, as these, from the mineral in them, scratch and impede the burnisher.

Detail Colouring.

HE reader will please bear in mind, that the main portion of these notes applies to XIII and XIV century illumination. Also that, as previously stated, beauty of colour depends greatly on setting out one colour, on a different one, as the *pale* blue stem or leaf on a *deep* purple, or sometimes gold, or even on a *dark* blue ground, or the pale pink or lilac leaf or stem, on blue or deep purple. But we never have a stem or leaf darker than the background.

STEMS.—These are invariably lilac or pink, or pale blue, the former made up by adding more white and gum to the purple already described, when it becomes thick and glossy like an enamel body colour; the latter (pale blue), by adding more white and very little gum to the smalt ground colour, or using cobalt blue and white. The thicker parts of stems with the fret work thereon, have one side shaded darker with a little purple or some smalt, and thus better relieve the minute white circles and dots; the zig-zag on outer edging of these parts is like the edging of strawberry or bramble leaves, sometimes sharp, sometimes rounded. All these stems have a white line on one side to heighten light, but this, and indeed all white finishings, should be put on after the black lining. The same processes are adopted for the XIV century ivy stems, which have no backgrounds.

LEAVES.—The above two colours (lilac and pale blue) are used also in some leaves, shaded darker with purple and sienna tint for the lilac, and smalt for the pale blue, either from centre or from points of leaf. Some leaves are painted with orange chrome, and others are coloured with emerald green and a little oxide of chromium mixed;

and some with white, all shaded darker in same way as for lilac and blue—the white being done with pale grey, blue, or pale sienna, (the white not being the mere paper or vellum surface, but solid colour.) Buds at ends of stems which are the primitive, simple, graceful outlines of the sprouting leaves in spring, are painted in scarlet or orange chrome, not laid on too freely;* some are green, others white very slightly shaded with grey, or pale yellow sienna. If scarlet or green, they are edged with white. To these buds are generally attached gold balls. In note at p. 16, is given description of how other leafage as in Plate VI. is treated.

The ivy leaf of the xiv century, as p. 63, is coloured and shaded in the same way as in the xiii century above described, only *gold* leaves occur in this style, which are well raised, and sparkle amid the bright shaded blue, the scarlet and lilac leaves, scarcely two of the same colour being near each other. The bud in this style takes an angular form as in letter 'H,' p. 53. The letter 'K,' p. 60, although some leaves are of a different form, is coloured on same principle as the 'H.' At the end of stems or branches is a kind of knot, and this is almost invariably scarlet.†

FIGURE, DRESSES OR DRAPERY.—The figure drawing of the xii century artists‡ is undoubtedly after the manner of the Etruscans and Greeks, which is plainly seen, as already stated, on vases and relics from Pompeii, &c. In xiii century work, less labour is bestowed upon

* The scarlets of many of the ancient MSS. have turned quite black, like black lead. A good permanent brilliant scarlet is much wanted.

† The stems or leaves in xv and later work are coloured naturally— and are consequently mostly green. The reader may think that because we have so much green in nature we should follow it for decoration. But it has been the very reverse in all good decorative art; and therefore little green was used in the early styles, as if we had quite enough of it in its own place ; but not enough of deep rich blues, purples, lilacs, scarlets or even gold.

‡ See Plate I.

figures and drapery, but at the same time they are treated with more artistic and a little more natural effect, as the artists advanced in knowledge and skill. The capitals and miniature subjects of xiii were not of such immense size as in xii century, and so required less manipulation, and fewer strokes in such small space to render the subject with proper effect. Therefore, while we have the features elaborately though coarsely painted, and the nails, joints, and veins of fingers and toes shown, with masses of flowing lines for drapery and hair, in the xii, these are confined in xiii century to almost mere outline. The whole arrangement is better; it is not the straight but the waving curly locks of hair, a few single strokes for the expression of face, and the general look of hands and feet without counting the fingers and toes. The whole figure is better disposed, as if the artist had thought of the structure of the body beneath the drapery. This is a point illuminators should attend to in figure drawing, particularly in sitting positions, and ask themselves if it would be rightly proportioned without the drapery. This method was strictly carried out by Michael Angelo, and all great artists practise this rule.

Some illuminators think that the vellum surface was left untouched for the colour of the faces and hands, but on close observation it will be found that white body colour was laid on; this should be done after burnishing the gold, and before the faces are delineated. The white may be rough, but if you lay a piece of satin paper on it when nearly dry and burnish it, or press it with a paper cutter, it will look much better. The features on this white surface are best done with the smallest hair pencil made, not a pen, as it will clog or scratch up the white: be careful to set the iris or eye-dot in its right place, the breadth of a hair's point may make all wrong. Put a touch of pale rose colour on the cheeks only—not the mouth, else the whole expression will be altered. This you would do well to try from some of the faces in illustrations, noting the totally different expression you obtain by making the mouth straighter, or placing the iris or dot at bottom instead of top of eyelid. The hair in XIII century work is not coloured, save with same white as for face. Of all parts of a picture, drapery is the most difficult to dispose gracefully and to draw effectively. The folds of drapery in XIII and XIV century styles are expressed by bold lines taking the place of the shadow that one fold or overlap casts beneath it; and one line more or less may spoil the effect. So difficult is this part for the beginner, that for a long time he cannot do better than copy most scrupulously the drapery drawing of these periods: and at the same time he should study this subject by sketching from nature when he sees graceful folds—in the full light cloak, for instance, as now worn; or even in the fling of a milk-girl's shawl over her right shoulder; or the hanging of curtains, &c. You cannot get much grace out of men's dress except in the Inverness or military cloak. Some good draperies, besides numerous useful details may be got from Mr. Edward W. Lane's " Manners and Customs of the Modern Egyptians," also from the photographs now so common, of sculpture from the French Cathedrals at Paris, Amiens, Rheims, Chartres, &c.

Drapery in XIII century illuminations was painted principally with the sacred colours, blue, purple, and scarlet, or compound tints of these. If the under dress was blue, the upper garment or cloak was red-purple, or visa versa; the inside lining when exposed, being scarlet or white: other parts requiring a different contrasting or lighter colour were painted by mixing the blue and purple with white, giving it by the former predominating, a violet-carmine tone, or by more of the purple what we may call an inky-pink tone, like the pinkish slates from Bangor. Blue, purple, scarlet and gold, were always employed as royal colours for the dresses of God, our Lord, David, and Kings, the Virgin Mary, and Joseph.* In xiv century, similar colours, but their modifications more heightened, were used, the drapery folds still being expressed by lines, but by slight shading in addition, as in p. 71. Dresses were sometimes richly diapered at this period as in the Berri Bible, &c. Here began a change—the drapery being treated like the drawing of Albert Durer. For xiv century work, you cannot get nobler examples than the wood engravings and etchings of this great master. If you can obtain the rare series of his "Passion of our Saviour," (some thirty-four prints) you might paint them in panels, such as in Plate IX., with varied ivy borders, as in Addl. 23,145, British Museum, and for a sacred task, or essay, compile a brief history of Christ's sufferings. In this you would be greatly assisted by using the "Tabular View of Gospel History," in the 'Paragraph' Bible, published by the Religious Tract Society, which may also be recommended for giving all the poetical parts of Scripture in a more poetical form or in verse, than in the general prose.

NIMBI.—The Nimbus, or circular glory round the heads of the three Divine persons in the Godhead were invariably gold, as the grandest colour that could be devoted to them. The Father and the Holy Spirit have it plain, unless they in some way personate Christ, when a cross is introduced as in Plate I. where God sits as the Alpha and Omega, and the "I am that I am," and yet personates Christ by having the cross. The Holy Spirit is represented by a dove, with golden nimbus round the head, sometimes with the cross also—the old illuminators seeming to have a clear idea of the equality of the holy three forming the one Godhead. On these points we would again refer the reader to Didron's "Iconography of Christian Art." The nimbi of saints are generally gold also, unless many occur in the same picture, when

* All drapery was edged with a white thin line within the black, like a hem or delicate embroidering.

they are painted blue, purple, scarlet, or greenish blue, much depending on the contrast needed against the background.

BUILDINGS.—These are of so many varieties in treatment and colour, that it would fill pages to describe all, but the best way to derive knowledge on these points is plainly to look at MSS. The architecture should be of the same date as the style of illumination you are engaged upon, but the colouring depends on the contrast requisite. It may be pale blue, or pink, with scarlet roof or turrets, with the battlements and courses of stone lined in thin black or white; the *outline* being firmer black. Windows and doors are coloured darker, or gold; in later styles, windows are in silver, quarried with a diamond pattern. Seats or thrones as in letter ' V ' sequel, may have two or three colours— a small part gold or green, or scarlet, rest blue, all depending again on colours different from dress of person seated thereon.

TREES.—Some good illustrations and useful information on trees, &c., we again remind the reader may be obtained in third volume of " Modern Painters." These are generally green in trunks and foliage, except in late XIV and XV century work. A better idea of treatment of trees in illumination of the early periods cannot be given than in piece of painted glass from Chartres Cathedral, Plate VIII., where two styles of foliage are shewn exactly similar to what occur in illuminated MSS. Another style of foliage is like large fir-cones set on the branches. Birds and scarlet apples are often introduced with good effect in some examples.

ROCKS.—Rocks seem to have been a puzzling matter to the early illuminators, and are attempted in their natural colours. One example is given in p. 71. They are often most skilfully painted in late work.

FOREGROUNDS.—Foregrounds were seldom attempted by XIII century artists, who made their figures stand on the lower bar of a capital as the ground; but later in XIV century, the ground was beautifully painted like soft velvety grass with delicate flowers scattered about.

RIVERS, THE SEA, AND WATERS were in both periods painted at first in emerald green with waving lines and fish drawn in black. Probably the artists never saw the sea, but they were not far wrong in making it green. Later it became bluer, streaked with white for the glitter of waves; fish, and occasionally queer objects (such as we have in our aquariums) swarm about. In some debased work, rivers were painted in dull silver.

DIAPERS. See pp. 50 and 51.

BLACK LINING also has been previously explained sufficiently. Only it may be added that black lining on blue, must, because of its deep colour, be very thick and gone over twice. The whole brilliancy of the drawing depends on the firm glossy black line, and till this is put on, you should not show your drawing except to those who can judge of the difference between a first rough coloured sketch, and what its finishing may effect. The straight black lines are best ruled in with the steel bow pen, which by the small screw you regulate to the thickness of line required. Clean the edges well inside and out with a slip of thick paper, and then put the ink in between the lips of the pen with a pen or brush, and rule in with small **T** square or straight edge. Always clean the pen well when laid aside.

WHITE LINE OR HAIR FINISHING.—The Chinese white in bottle is preferable to any other for constant use. On opening it, half a teaspoonful of pure white spirits may be poured in, and thus it will always keep moist; and it should be always immediately well corked after taking out any quantity. Mix the white in a small saucer, to the consistency of cream, and then use it; and as it dries and thickens in about an hour, add a drop more water. The only steel pen that can be used sometimes, is Perry's lithographic pen, of such thin metal that you can cut a pen for yourself with a knife. Other steel pens are too hard, and scrape up colour beneath. Crowquills, although ever so fine, clog up the colour and become too soft. So the best method is, at once to adopt the smallest brush made, and practice with it. It certainly requires some delicate, firm skill to put in the fret work, fine veins and dots. It may be thought very wonderful work, but becomes easy if you will give your mind to it. The flowing fret, as on long main straight stroke of ' P,' Plate III., is done by the single equal waving line first, and the leaflet springing therefrom after. Much depends on the floweret dots being intense and brilliant, and they may therefore require to be gone twice over with white. It will be observed that all white finishings are on the most minute scale compared to other details, and for this reason, that if the whites were in larger mass, they would completely kill or destroy the whole page of colour. The straight lines may be ruled in with the bow pen as for the black lines, only be careful that the line is thin, and that the pen is cleaned after black ink. All white finishing on purple grounds, and lilac or pink parts, such as stems and leaves, must be twice gone over, as the first wash of white always sinks into these two colours, and is almost lost. The blue, &c., do not require white to be repeated.

Design.

N this subject a few words may be added to follow up Mr. Goodwin's remarks, and may also prove useful to the reader. First, dispossess yourself of the idea that good ' design' must be entirely original. Take even the best designs, and it will be found that many of their best component parts can be traced to pre-existing examples. This will ever be so, as long as our eyes are

open and habituated to good and beautiful impressions.
Therefore in copying good examples, you form your
taste, and are likelier to produce good work. A blind
person may design something original, but it would most
likely be an abortion—it is *blind* people, who will not
see, but through their own conceits, that produce the
stupidest things.

The most profitable lesson towards design, is:—1st. to
have a thorough *conception* of the subject to be treated ;
2nd. to add thereto *invention;* and 3rd. *completion*—seeing that all parts harmonize, and that the whole is properly
finished. In Part III. the first idea is already treated of
sufficiently. We may only add that a good pencil sketch
is sufficient without inking it in, which is a waste of time,
and besides the ink is apt to wash up and mingle with
the pure colours. For our second point—invention—the
beginner, with a view to gradual improvement, should
at first copy a part of an illuminated page, the mere border for instance, and the miniature capital or great initials
(according to word required to begin a certain passage)
from another page of same MSS., so as to ensure same
style, and unite it to border, as suggested at pp. 35 and
36, Part III. After sketching it in, compare its arrangement or composition with similar examples even in
other MSS. but of the same style or date. In your next
attempt you might alter the details a little, such as the
setting of the leaves. Although variety is the soul or
spirit of all Gothic art, be cautious at first as a beginner,
and always endeavour to see an authority for what you
do. The third point—*completion*—is the carrying out
of the details in colour, and the full and final develope-
ment of all finishings.

The next advanced step towards design is, when you
see a good thing in drawing and colour that strikes you
as adapted for illumination, to take note thereof, and
try something like it. Much has been said in previous
parts of this Manual as to the close connection between

architecture, glass-painting, and illumination. Of the first some examples are shown in illustrations ; of the second, one is given in Plate VIII. from the lower part of the painted window of Saint Egyptiaca in clerestory of Chartres Cathedral,* which shows the same treatment in figure drawing, in drapery, and foliage as in XIII century illumination ; and which, if the student colours like the original, save the white dress of the priest, of which there would be too much in illumination (but not in glass-painting) it will afford a pleasing lesson. The dotted bordering is only inserted to make it a complete picture, for the window has the full length figure of the saint above, with canopy in arch head. Assuming then that it is good, and from its interest that you would like to make it form part of a small missal, let us see how we may use it for illumination. It is XIII century work, and we find something like its border in the ' I ' from Vegetius in Plate III, repeatedly in border ' I's' from the Bible in Advocates' Library at Edinburgh, and in other MSS. all uniformly round as in our design. Generally all painted windows are *read* from the lower part, but in our illumination this must be altered. Egyptiaca comes to the priest first, so it should be the leading subject, and as a very useful

---

* This interesting window is referred to by the late Mrs. Jamieson in her " Sacred or Legendary Art," p. 389, and is thus described by the Abbe Bulteau in his ' Description de la Cathédrale de Chartres.' " In the lower tablet we see Marie (l'Egyptienne) upon her knees, with the mantle of Zozime ; she relates the history of her life ; Saint Zozime is seated and listens with earnest attention." In the second tablet is painted the burial of Sainte Marie l'Egyptienne ; which is related thus by Jacques de Voragine : " Whilst the old man was endeavouring to dig out the earth, which he was too feeble to do, there came to him a very gentle lion, to which he said, ' This Saint has commanded me to bury her, and I am unable to dig the grave, because I am so old, and have not the necessary implements. Do thou, therefore, break the earth, and scrape out sufficient that we may inter the body.' And the lion commenced assiduously to work, and made a proper grave ; and after the body was interred, it went on its way as gentle as a lamb ; and the old man returned to his monastery glorifying God."

H

lesson in increasing from small or larger drawing the lower
compartment of window will therefore require to be enlarged; and her burial as the next event, will have to be
reduced for the capital, within lines as shown. To show
the license the ancient artists took, rather than crowd
and spoil any picture—the lion boldly infringes on the
border, and you may do the same. Here is our first rule
—conception—shown by re-arranging and borrowing. The
second—invention—is shown by such things as having
eight whirls of stem on one side and seven on the
other, dragons unlike, one with legs, one with wings,
&c.; and our third—completion—in making the whole
harmonize (mainly by colour), by inserting gold balls
to balance effects, of which one more or less may destroy
all; and lining in with black and white. To finish even
our piece of glass as an illumination, gold balls as dotted
are required, else the spaces would look blank. Yet the
suggestive design is not altogether right, although very
beautiful when coloured; it wants the wild outline liveliness
of the Gothic spirit, seems too like the Renaissance; but
is given as a specimen to explain also the limits we must
observe. A livelier, and as you will see, nearer approach
to our window example, is the Hebrew border, which is
carried round square most beautifully in another page of
same Hebrew MS.

At p. 14, it is stated that more was to be said on the
fine initial ' B,' p. 41, and well we may, for in nine out
of every ten Psalters, you will find the subject treated in
a similar manner. In it is shown great design. The
two leading features in David's life were, 1st. his slaying
Goliath, which led in God's arrangement to 2nd. his being
made and chosen king; so in the ' B ' of the Beatus of
the Psalms, these are always prominently expressed, with
the addition of the harp, from his love of music. Then
there is some inventive meaning in the lion's head at the
middle bar of the ' B,' as the trophy of his first field
exploit while keeping "those few sheep in the wilder-

ness," where he practised with his sling and his bow, with the wonderful eye and arm God gave him, besides the fine toned ear to practice music to words of his own poetry, on a rude harp, perhaps of his own construction.

The student will get suggestions and ideas everywhere for design, diapers from fine old-fashioned damask table-cloths, from Scotch tartans even, not unlike some in Berri Bible, and Addl. 23,145, &c., British Museum; or griffins from good old china, like the one in Plate IX., &c., or by picking up a twig, and cutting off two branch-lets from its lower stalk, like the letter ' Y,' and so get that form of knot generally at ends of long stems, from which curved stems spring, as in Plate IX., letters ' H ' and ' K,' &c., or the rounded knob-knot from a twig also as in Plate VII.; these being conventionalized from the true natural forms. For scrolls, we might simply enlarge from small borders. But for confusing the reader, we might have gone farther into consideration of the art in its application to work on a large scale, such as the proper treatment of the Ten Commandments, the Lord's Prayer, &c., as required in our churches.

In conclusion, we may ask if this age of steam, and machinery, accomplishes so much for us, makes so many books for us, might it not be reasonable that one should make one book for himself? So much is done by machinery now, that to keep pace with it, men are becoming mere machines themselves, and despise any manly handywork. But think if we could attempt any nobler or more beneficial task than to copy and decorate a certain portion daily of the Book of Life by our own hand. Can we ever make enough of it, or esteem it too highly; should we slight any means to make us think more upon its words; and could any one dare to say the time would be mis-spent, although we have tons of it printed to our hand? To employ the art to such noble purpose, or on any worthy book or piece of work, by penning it with our own hand, is not a mere monkish pursuit, nor a retrograde step of the so-called 'dark ages.' What greater good could parents think of, than that it should be a rule of their house—that their children on coming to the age of fourteen, should devote some time every morning early to copying a portion of scripture, and make it a law of the first importance, to which every other duty shall bow; and by this Book—this Bread of Life, nourish and train them up in habits of regularity, of patience, reflection and goodness? No one can question the importance of such a habit for a moment, not even by measuring it with the most utilarian demands of the age. These reasons and the great advantages gained in every sense from such a pursuit, might well be our greatest plea for the revival of the art; and on its broadest basis we cannot do better than repeat the truism with which we commenced the first edition of this Manual, viz:—"The Art of Illumination has this rare advantage, that to the contemplative quietude of the library it unites the fervent love and pursuit of nature which form the enthusiasm of the studio. Thus it is at once disciplinary and delightful, and tends, even as an accomplishment, to strengthen those qualities of patience, thoughtfulness and delicacy, which shed so salutary an influence upon our daily life."